Improving Health-Care Management in the Workplace

Work in America Institute's National Policy Studies

With the publication in fall 1985 of Work in America Institute's newest national policy study, *Improving Health-Care Management in the Workplace,* the Institute will have completed its sixth major study since January 1978. Previous studies are:

Job Strategies for Urban Youth: Sixteen Pilot Programs for Action
The Future of Older Workers in America
New Work Schedules for a Changing Society
Productivity through Work Innovations
Employment Security in a Free Economy

A companion volume to *Improving Health-Care Management in the Workplace* is *Breakthroughs in Health-Care Management: Employer and Union Initiatives* by Victoria George and William E. Hembree, another in the Pergamon Press/Work in America Institute Series.

All of the above books are available from Pergamon Press, Inc.

About Work in America Institute

Work in America Institute, Inc., a nonprofit, nonpartisan organization, was founded in 1975 to advance productivity and the quality of working life. It has a broad base of support from business, unions, government agencies, universities, and foundations, as reflected in its board of directors, academic advisory committee, and roster of sponsoring organizations.

Through a series of policy studies, education and training programs, an extensive information resource, and a broad range of publications, the Institute has focused on the greater effectiveness of organizations through the improved use of human resources.

Improving Health-Care Management in the Workplace

Work in America Institute's National Policy Studies

With the publication in fall 1985 of Work in America Institute's newest national policy study, *Improving Health-Care Management in the Workplace,* the Institute will have completed its sixth major study since January 1978. Previous studies are:

> *Job Strategies for Urban Youth: Sixteen Pilot Programs for Action*
> *The Future of Older Workers in America*
> *New Work Schedules for a Changing Society*
> *Productivity through Work Innovations*
> *Employment Security in a Free Economy*

A companion volume to *Improving Health-Care Management in the Workplace* is *Breakthroughs in Health-Care Management: Employer and Union Initiatives* by Victoria George and William E. Hembree, another in the Pergamon Press/Work in America Institute Series.

All of the above books are available from Pergamon Press, Inc.

About Work in America Institute

Work in America Institute, Inc., a nonprofit, nonpartisan organization, was founded in 1975 to advance productivity and the quality of working life. It has a broad base of support from business, unions, government agencies, universities, and foundations, as reflected in its board of directors, academic advisory committee, and roster of sponsoring organizations.

Through a series of policy studies, education and training programs, an extensive information resource, and a broad range of publications, the Institute has focused on the greater effectiveness of organizations through the improved use of human resources.

Improving Health-Care Management in the Workplace

A Work in America Institute Policy Study

Directed by Jerome M. Rosow, President
and Robert Zager, Vice President
for Policy Studies and Technical Assistance

and Ruth S. Hanft, Special Consultant

Pergamon Press
New York • Oxford • Toronto • Sydney • Frankfurt

Pergamon Press Offices:

U.S.A. Pergamon Press Inc., Maxwell House, Fairview Park, Elmsford, New York 10523, U.S.A.

U.K. Pergamon Press Ltd., Headington Hill Hall, Oxford OX3 0BW, England

CANADA Pergamon Press Canada Ltd., Suite 104, 150 Consumers Road, Willowdale, Ontario M2J 1P9, Canada

AUSTRALIA Pergamon Press (Aust.) Pty. Ltd., P.O. Box 544, Potts Point, NSW 2011, Australia

FEDERAL REPUBLIC OF GERMANY Pergamon Press GmbH, Hammerweg 6, D-6242 Kronberg-Taunus, Federal Republic of Germany

Copyright © 1985 Work in America Institute

The publications produced as part of this project were prepared under grants from the Atlantic Richfield Foundation, General Electric, the Metropolitan Life Foundation, Pfizer Inc., and the United Auto Workers–General Motors Joint Committee on Health Care.

Library of Congress Cataloging in Publication Data

Work in America Institute.
    Improving health-care management in the workplace.

    (A Work in America Institute policy study)
    Bibliography: p.
    1. Occupational health services. 2. Health
services administration. 3. Health planning.
I. Rosow, Jerome M. II. Zager, Robert.
III. Hanft, Ruth S., 1929-   . IV. Title. V. Series
RC968.W66   1985   658.3'82   85-19238
ISBN 0-08-032797-4

All rights reserved. No part of this publication may be reproduced, stored in a retrieval system or transmitted in any form or by any means: electronic, electrostatic, magnetic tape, mechanical, photocopying, recording or otherwise, without permission in writing from the publishers.

Printed in the United States of America

CONTENTS

Preface	xi
1. Health-Care Management in the Workplace: An Overview	1
2. Joint Action: Employers/Employees/Unions	18
3. Designing Health-Care Benefits	34
4. Managing Health-Care Benefits	53
5. Work-Site Wellness Programs: Fad or Future?	75
6. Health Maintenance Organizations	93
7. Corporate and Union Strategies for Health Care	113
8. Reducing Excess Hospital Capacity	129
9. Controlling the Use of Medical Technology	142
Glossary	151
Index	155

NATIONAL ADVISORY COMMITTEE
IMPROVING HEALTH-CARE MANAGEMENT IN THE WORKPLACE

+Richard Allen
Vice President
Personnel Research & Planning
Pfizer, Inc.

*Theodore Allison
Vice President
Government and Industry Relations
Metropolitan Life Insurance Co.

*David Beier
Administrator, Benefit Plans Section
GM Department
United Auto Workers

+Theodore Bernstein
Director, Benefit Funds Department
International Ladies' Garment
  Workers' Union

*Joseph Bilich
Second Vice President, Personnel
The Travelers Companies

Warren Billings
Manager, Benefit Planning & Analysis
AT&T

Irving Bluestone
University Professor of Labor Studies
Wayne State University

Philip Briggs
Executive Vice President
Metropolitan Life Insurance Co.

Dr. John M. Burns
Corporate Director
Occupational/Environmental Health
Honeywell, Inc.

Dr. Philip Caper
Visiting Professor
Dartmouth Medical School

*Nelson Carpenter
Associate Director
Governor's Office of Employee
  Relations
New York State

John C. Carroll
Executive Vice President
Communications Workers of America

David C. Collier
Former Vice President & Group
  Executive
General Motors Corporation

Donald F. Ephlin
Vice President
Director, UAW/GM Department
United Auto Workers

*Judith Fleming
Vice President, Employee Benefits
Lincoln National Life

Marlene Fryer, R.N.
Occupational Health Nurse
Olin Corporation

+Melvin Glasser
Director, Committee for National
  Health Insurance

Arnold Glassman
Manager, Benefits, Communications
  and Programs
Atlantic Richfield Company

+Dr. Robert Greifinger
Medical Director and Vice President
Westchester Community Health Plan

*Jim Gutowski
Deputy Director
Division of Management/Confidential
  Affairs
Governor's Office of Employee
  Relations
New York State

Thomas F. Hartnett
Director
Governor's Office of Employee
  Relations
New York State

Stanley Hill
Associate Director
District Council #37
American Federation of State,
  County, and Municipal Employees

*John Hillins
Director of Executive Compensation
Honeywell, Inc.

Elizabeth L. Hoke
President
New York State Public Employees Federation, AFL-CIO

Ronald Hurst
Manager, Health Care Planning
Caterpillar Tractor Co.

Karen Ignagni
Assistant Director, Department of Occupational Safety, Health and Security
AFL-CIO

Judith Miller Jones
Director
National Health Policy Forum

*Dr. Geoffrey Kane
Director, Health Services
Special Programs
Blue Cross-Blue Shield

Richard Kennedy
Staff Executive and Manager
Health Care Management Program
General Electric Company

Dr. Clark Kerr
President Emeritus
University of California

*Howard Kline
Assistant Director
Insurance, Pension and SUB Department
United Steelworkers of America

Dr. William Mayer
Assistant Secretary of Defense (Health Affairs)
U.S. Department of Defense

*Laird Miller
Manager of Health Care Management
Health and Environmental Resources
Honeywell, Inc.

Jim Mortimer
President
Midwest Business Group on Health

+Patricia Nazemetz
Benefits Operation Manager
Employee Benefits
Xerox Corporation

*Candice Owley
Vice President
American Federation of Teachers

+Dr. Peter Rogatz
Vice President for Medical Affairs
Visiting Nurse Service of New York

Dr. Lindon E. Saline
Former Staff Executive and Manager
Health Care Management Program
General Electric Company

*Gerald M. Shea
Health Care Industry Coordinator
Service Employees International Union

John J. Sweeney
International President
Service Employees International Union

Walter Trosin
Vice President, Human Resources
Merck & Co., Inc.

+Dr. Leon Warshaw
Executive Director
New York Business Group on Health

Dr. David West
Director of Special Projects
Hospital Corporation of America

The following served on the Design Committee only:

Stanley Brezenoff
President
Health and Hospitals Corporation
New York City

Dr. Richard Stone
Medical Director, Research
AT&T

Diana Walsh
Director, Program Evaluation
Health Policy Institute

---

* Associate Member
+ Also Member of Design Committee

After two decades of standing by while the U.S. health-care system was spinning into hyperinflation, the consumers and purchasers of health care have finally introduced measures to achieve some control. Employers and unions, in particular, are learning to manage so that each dollar buys a more favorable mix of price, quality, and accessibility of health care. However, there is a great distance still to be traveled. *Improving Health-Care Management in the Workplace* describes the turnabout in health care, outlines how much farther we need to go despite the accomplishments of the last few years, and identifies a number of practical and effective programs for getting there. On behalf of the board of Work in America Institute, I commend this report to everyone who uses or pays for health care: in short, to everyone.

CLARK KERR
Chairman of the Board
Work in America Institute, Inc.

PREFACE

Uncontrolled health-care costs take money out of the pockets of employers and employees alike. Although employers pay the lion's share, a large part of that payment would otherwise have been available for other fringe benefits and for wages and salaries. With so much at stake for both sides, one might expect this to be an ideal area for joint action, but this has not been the case, with several notable exceptions. This report outlines several of these exceptions—instances in which employers and unions have taken joint action to improve the management of health care in the workplace—and proposes means by which other employers, employees, and unions can achieve comparable results.

The report—and a companion book of cases, *Breakthroughs in Health-Care Management: Employer and Union Initiatives* by Victoria George and William E. Hembree, to appear in spring 1986—stem from an 18-month national policy study conducted by the Institute from early 1984 to mid-1985. The need for the study was first conceived by Frank Doyle, senior vice president, corporate relations staff, the General Electric Company, and General Electric subsequently took the lead in providing major financial and professional support for the study. Additional support of both kinds was provided by the Atlantic Richfield Foundation, the Metropolitan Life Foundation, Pfizer Inc., and the United Auto Workers-General Motors Joint Committee on Health Care.

Work in America Institute was fortunate in assembling for this policy study a national advisory committee of diverse and practical talents, whose spirited debates illuminated the issues we had set out to understand. The committee met four times over the course of the study, reviewing on each occasion four background papers prepared by experts in the health-care field. We were particularly fortunate in obtaining the assistance of Ruth S. Hanft as our principal consultant. A former deputy assistant secretary for health policy in the U.S. Department of Health, Education, and Welfare, Ms. Hanft is now a consultant in the health-care field, with emphasis on delivery, financing, manpower needs, and academic health programs. In addition to writing four of the background papers, she drafted a large part of the report and rendered valuable advice on the drafting of other parts.

Although the comments of our sponsors, advisory committee members, and experts have strongly influenced our thinking, Work in America Institute alone is responsible for the findings and the 28 recommendations made in the report.

The authors and topics of background papers were as follows:

☐ Richard Allen—*Health Care Cost Management at the Worksite: The Pfizer Approach*

☐ Philip Caper, M.D.—*Health Care Cost Containment and Patterns of Medical Practice: The Role of Business*

☐ Robert B. Greifinger, M.D.—*Building Physician Alliances for Cost Containment*

☐ Ruth Hanft—*Worksite Wellness Programs: Fad or Future?; Alternatives to Hospital Care; Employers and HMOs; Data and Management Information for Monitoring Effective Use of Health Care Services and Cost Containment*

☐ Daniel F. Hanley, M.D., and David N. Soule—*The Maine Medical Assessment Program: Informed Inquiry by Practicing Maine Physicians into Common Medical and Surgical Treatments*

☐ William E. Hembree—*Joint Labor and Management Efforts to Control Employee Health Costs*

☐ James R. Kimmey, M.D.—*Dealing with Excess Hospital Capacity: Methods and Consequences*

☐ D.N. Logsdon, M.D.—*Preventive Services and the Management of Health Care*

☐ Francis D. Moore, M.D.—*Asymmetry in Medical Resource Utilization*

☐ Jim Mortimer/Midwest Business Group on Health—*Model Competitive Health Care Purchasing System*

☐ Joseph Newhouse, Ph.D.—*Summary of the Rand Health Insurance Experiment*

☐ Lindon E. Saline, Ph.D.—*An Overview of The Business Roundtable Health Initiatives*

☐ David West, Ph.D.—*Pre-Admission Certification: A Golden Opportunity to Start Managing Health Care*

We are particularly appreciative of the contributions of Robert Zager, Work in America Institute's vice president for policy studies, who was responsible for the initial planning and overall direction of this policy study. We wish also to thank the members of the publications and policy study staffs of the Institute, particularly Beatrice Walfish, editorial director; Frances Harte, director of marketing and public affairs; Joan White, senior production assistant; Stephanie McDowell, production assistant; Sandi Frank, editorial assistant; and Virginia Lentini, assistant to the vice president for policy studies. Without their best efforts, this report could not have been published.

> Jerome M. Rosow
> President
> Work in America Institute, Inc.

1. HEALTH-CARE MANAGEMENT IN THE WORKPLACE: AN OVERVIEW

Discussions of health care in the United States reverberate with anger and accusations, often obscuring the good points of the system. The predominant faults are not the work of any one group—politicians, doctors, hospital trustees and administrators, insurance companies, employers, employees, or unions—although each of these shares some portion of the blame. The faults are due to good intentions gone awry, ideological blinders, impatience for action, inattention to the foreseeable consequences of action. We cannot alter the parties involved, but we can—with patience and attention—alter the circumstances under which health-care services are performed.

This study does not propose any wholesale change of the system; it advocates neither an all-regulated nor an all-free-market approach. There is so much room for improvement in the present mixed system that no one really can tell whether either of those alternatives might be better.

Neither does the study presume to instruct physicians, hospitals, and other providers on how to run their businesses. Work in America Institute's audience consists of individuals and organizations who *purchase* health services: employers, unions, employees, dependents, retirees. As they become more cost-effective users, the providers will have to become more cost-effective too.

Sowing the Wind

Historians will view the 1980s as a watershed in the history of American health care. Major, rapid changes are affecting all aspects of the health sector, particularly the financing and organization of care.

2 Improving Health-Care Management in the Workplace

The last great revolution in health care—which began in 1910 with the publication of Abraham Flexner's report advocating fundamental changes in medical education—changed the scientific basis of health care, stimulated rapid technological advances, and set the pattern for health care for 70 years. The reform of medical education and succeeding technological advances dramatically altered physicians' roles, ability to intervene in illness, practice patterns, power, and autonomy in society. Hospitals, which had been warehouses for the dying and the poor, became laboratories for a new type of physician and, ultimately, a major influence on technology diffusion and health-care costs.

The growing recognition of the value of physicians' skills during the last seven decades made the physician the focal point for decisions about most aspects and sites of health care.[1] The physician became the key determinant of demand for services and consequently for expenditures. This central role was enhanced by the development of third-party payments, which reduced the cost of care at the time of service and assured payment for the physician and for services ordered by the physician. Physicians set their own fees, which were reimbursed, as requested, by third-party payers. Reimbursement recognized virtually all costs, particularly for services provided in hospitals and for most tests and procedures.

The methods of payment—retroactive, reasonable costs for hospitals and fee-for-service payments to physicians—by their very nature stimulated the volume and intensity of service and the expansion of capital and technology. Physicians incurred few costs for treatment decisions, and hence they had no incentive to withhold even marginally useful diagnosis or treatment modalities. Until recently, the hospital had no incentive to constrain these decisions either.

In addition, the consumer, inhibited by lack of technical knowledge, had little direct economic stake because of the rapid development of employment-based private health insurance after World War II.[2] Health-insurance benefits became a major fringe benefit demand in union-management negotiations, spreading rapidly from large to small employers and to nonunion industry. The scope of health insurance expanded from coverage of hospital care to surgery to out-of-hospital physician services and, most recently to services such as dental care and home health benefits.

Private insurance coverage, however, because of cost, did not reach the population out of the labor force. Then, in 1965, after two decades of unsuccessful efforts, the federal government enacted the Medicare program for the elderly and the Medicaid program for the poor. Medicare was later extended to the disabled beneficiaries of Social Security and to the population with end-stage renal disease.

The reimbursement methods selected for Medicare—retroactive reasonable costs for hospitals and reasonable charge payments for physicians—influenced methods of payment in the private sector and were inflationary. Other factors on the supply side of the health industry also contributed to these inflationary trends. Congress, concerned about the accessibility of hospital care, passed the Hill-Burton Act in 1948. The Hill-Burton program was designed to increase national hospital capacity and to develop hospitals in underserved areas. This legislation, the subsequent development and use of tax-free bonds, and the recognition of interest and depreciation payments as reimbursable under private and public health insurance led to a dramatic expansion of hospital capacity.

In the early 1960s, during the protracted legislative debate on Medicare, public-policy analysts and legislators feared that there would be an inadequate supply of physicians to meet new demands for care if Medicare were enacted. The number of medical schools and physician-to-population ratios had declined precipitously after the Flexnerian reform of medical education and, though demand for medical services increased with the advent of health insurance, physician growth lagged. Physicians were deserting small towns, rural areas, and inner cities. In the 1950s, foreign medical graduates were given immigration priorities to help fill the gap. The American Medical Association fought direct federal funding of medical education until the 1960s.

In the 1960s and 1970s, the federal government began to directly support medical education to encourage expansion in the number of schools and enrollment levels in existing schools, first through construction and institutional grants, and then through capitation (per-capita payments) and bonuses for increased enrollment.[3] States expanded the schools within their state university systems and added new schools. By the 1980s, the number of medical schools had grown from 80 to 127, and enrollment had doubled. Osteopathic schools grew from 6 to 15. Similar expansion occurred in the other health professions and in the development of health paraprofessionals to substitute for physicians.

4 Improving Health-Care Management in the Workplace

Medical education is a long process, seven years at the minimum, from entry into medical school to practice. Although the Graduate Medical Education National Advisory Committee Report of 1980 (GMENAC)[4] signaled concern about an emerging surplus of physicians, enrollment increases continued until 1982-83.[5] Furthermore, an estimated 3,000 to 4,000 U.S. citizens are studying abroad each year. The increased supply will not peak until well into the 1990s, but the effects are already beginning to be observed.

Reaping the Whirlwind

The results of the scientific revolution, the expansion of health

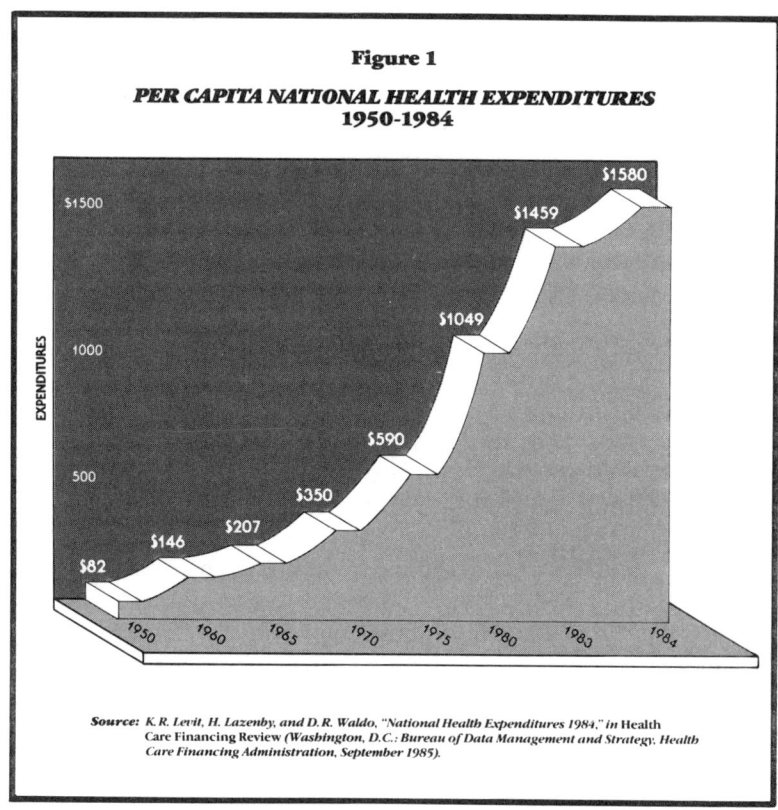

Figure 1
PER CAPITA NATIONAL HEALTH EXPENDITURES 1950-1984

Source: K.R. Levit, H. Lazenby, and D.R. Waldo, "National Health Expenditures 1984," in Health Care Financing Review *(Washington, D.C.: Bureau of Data Management and Strategy, Health Care Financing Administration, September 1985).*

insurance, capital expansion of hospitals, the increased supply of physicians and other health professionals, and increased consumer demand were rapidly rising costs and a health-care inflation that was double or triple the general economic inflationary trends during the last two decades.

Health-care expenditures in 1983 were $387.4 billion and constituted 10.6 percent of the gross national product, a higher percent of GNP than in virtually any other industrialized nation. In 1965, prior to the implementation of Medicare and Medicaid, these expenditures were $41.9 billion, or 6.1 percent of GNP. However, the percent of GNP spent on health began to grow very rapidly after 1965. Personal health-care expenditures have been growing at an average rate of 12.8 percent per year since 1965, with the lowest annual rate of increase, 9.1 percent, occurring between 1983 and 1984. Over 60 percent of the increase can be attributed to prices during this period, 8 percent to population growth, and 31 percent to intensity (technology and changes in use and kind of services). Per capita expenditures septupled between 1965 and 1984 (see fig. 1).

The sources of funding for health-care services have shifted dramatically. In 1965, the private sector was responsible for 78 percent of personal expenditures; the public sector, for 22 percent. In 1984, the private-sector share declined to 60 percent, and the public sector rose to almost 40 percent (see fig. 2). However, beginning in 1982, public-sector expenditures had started to decline.

Until 1965, direct out-of-pocket payments by consumers for personal health-care expenditures constituted more than half of all payments, private health insurance accounted for a quarter of the total, and the public sector was responsible for slightly more than 20 percent. In 1984, direct payments declined to 27.9 percent, private health insurance rose to 31.3 percent, and government payments were almost 40 percent.

The distribution of expenditures by type of service also has changed, with the percent spent for hospital care rising, with a recent slight decline in the proportion for physicians (but now increasing), and major increases in nursing-home care.[6]

Public Policy Changes

The rapid inflation in health-care costs (see fig. 3) led to numerous attempts by the public sector to constrain Medicare and Medicaid expenditures.

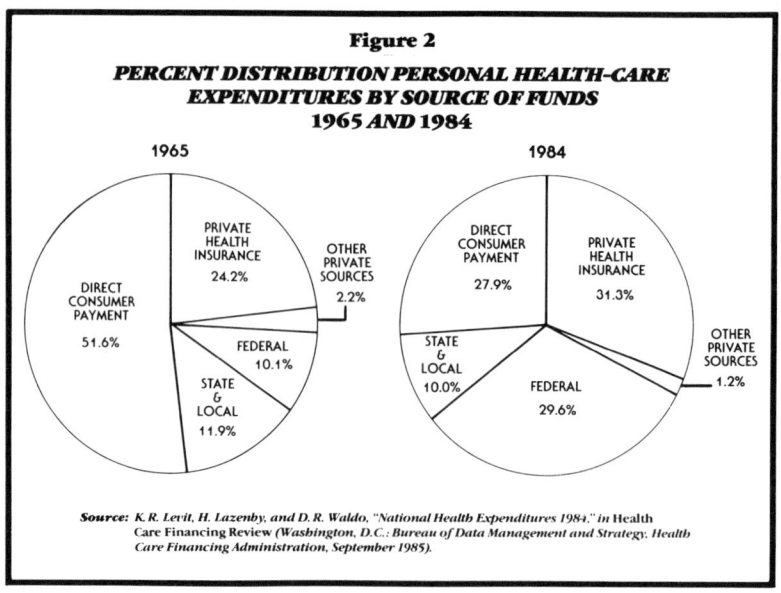

Source: K.R. Levit, H. Lazenby, and D.R. Waldo, "National Health Expenditures 1984," in Health Care Financing Review (Washington, D.C.: Bureau of Data Management and Strategy, Health Care Financing Administration, September 1985).

Prior to Medicare, hospitals were paid through a number of different arrangements: costs with discounts, costs plus a factor for capital expansion, charges, negotiated rates, and philanthropic contributions.[7]

When Medicare was enacted, the decision was made to pay hospitals on a retroactive reasonable cost basis. The consequences of payment of costs, whatever the costs, are obvious—incentives to provide more services, add equipment, hire personnel, and so on, since all costs are covered.[8]

These incentives became apparent soon after enactment of Medicare. Over time, numerous efforts were made to contain inflation in Medicare hospital costs, including:

☐ Limits on routine hospital costs[9]
☐ The economic stabilization program of the Nixon and Ford administrations
☐ Limits on ancillary costs[10]

Major changes were enacted in the 1982 Tax Equity and Fiscal Responsibility Act and the 1983 Social Security Amendments. These legislative changes radically alter the basis for payments to

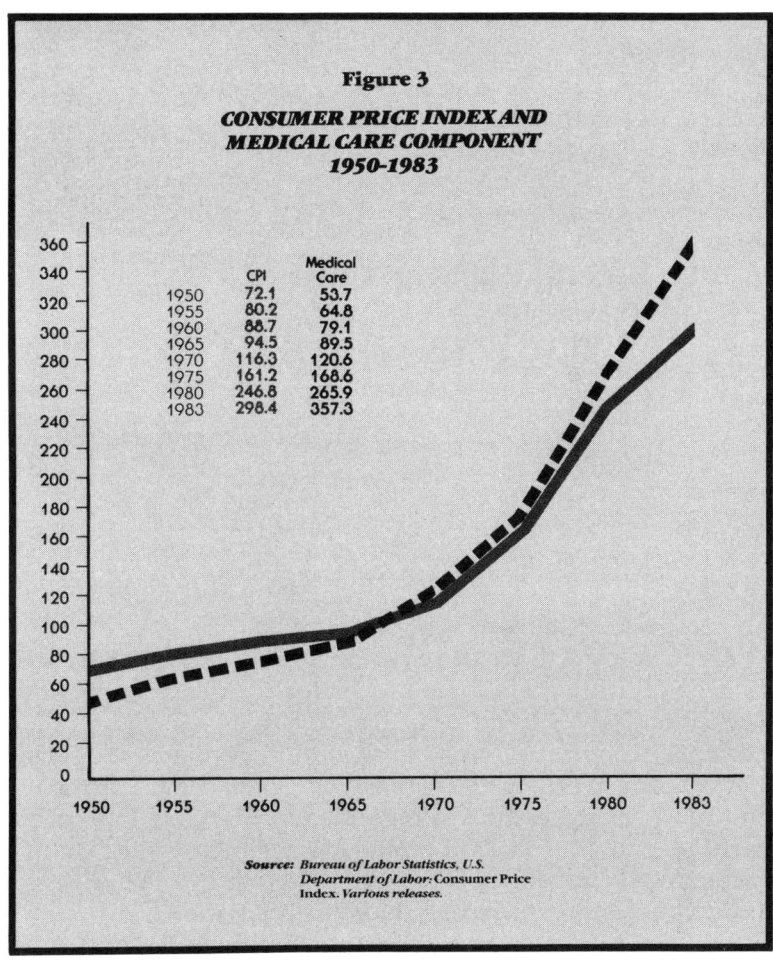

hospitals and all of the prior incentives in hospital reimbursement. The new method of payment is being phased in, and at the end of the period all acute-care general hospitals will be paid on the basis of national average rates for 468 diagnostic related groups (DRGs). Hospitals whose costs exceed these rates will lose revenue; hospitals whose costs are below these rates will profit. However, DRGs do not measure differences in the severity of illness within the diagnosis.[11]

This change is having a profound effect on the dynamics between physicians and hospital administrators in the following ways:[12]

☐ In the past, physicians had almost unlimited decision-making power (except for utilization review) on admissions, lengths of stay, and selection of ancillary services. Under the new reimbursement, patterns of physicians' practice that raise costs above the average will have to be reviewed if hospitals are to remain fiscally viable.

☐ The incentives in DRG reimbursement will:
—Shorten lengths of stay
—Encourage the use of extended-care facilities and home-health agencies
—Increase admissions
—Substitute out-of-hospital services whenever possible, such as preadmission diagnostic tests, day surgery, and so on
—Control the use of ancillary services in the hospital
—Ensure the treatment of less-ill patients

Reimbursement controls have not yet been placed on sites of care other than hospitals and nursing homes, and the impact of control on hospital costs is not yet known.

Changes in methods of payment to physicians are also occurring, and further changes are anticipated. The federal government is encouraging Medicare recipients to enroll in HMOs. As HMOs continue to grow and develop, capitation payments and salaried physicians will become increasingly common. HMOs provide the full range of inpatient and outpatient services, and physicians' remuneration is dependent on the balance between the use of inpatient and outpatient care. Thus, participating physicians in HMOs have an incentive to reduce hospitalization.

In addition, Medicare legislation required studies of the feasibility of paying physicians on the basis of diagnostic related groups. While this approach may not be feasible for all physician services, it may be the correct approach to surgery and in-hospital medical care. Discussions are also being held about the possibility of substituting fee schedules for the payment of "usual, customary, and reasonable charges." This approach would require physicians to accept the rate assigned by Medicare as full payment and bill the patient only for the coinsurance, the percentage of the bill which Medicare requires the patient to pay. In the interim, the federal government has frozen physicians' payments for 15 months.

Private-Sector Efforts

The continuing inflation in health-care costs has had equally profound effects on the ability of industry and labor to maintain their health-benefits programs. Health-benefits costs have risen faster than wages and other fringe benefits. Health-care costs paid by industry influence the price of products, the competitive position of U.S. industry internationally, and unemployment problems. However, until the early 1980s, with few exceptions, industry and labor made little effort to manage health benefits or to influence the use and costs of services.

In the last five years, health-care-cost coalitions have developed at local and regional levels. These coalitions, stimulated by industry and labor, take numerous forms. Some include both purchasers and providers, while others are limited to employers. Still others include both employers and unions. Their purpose is education and health-care management.

For a number of years, a major debate has been underway in the health-care policy field between proponents of increasing competition to create an economic market in health care and those who believe that an economic market is not viable. The latter group believes that health is an entitlement and that regulation is the only way to control costs.[13] The truth is probably somewhere in between.

The increase in the supply of physicians, the entry of venture capital into the health field, and the active intervention of management and labor have stimulated a competitive environment and the growth of market forces. Evidence of this growth includes:

☐ The expansion of HMOs, and their new ability to attract physician staff, due to the increased supply of physicians. One of the major barriers to the growth of HMOs in the past was the difficulty of recruiting and retaining physicians.

☐ The development of new types of services, such as variations on physicians' office practices and/or the unbundling of hospital services:
—Urgent-care centers—substitutes for physicians' offices and routine care in hospital emergency rooms
—Surgicenters for day surgery
—Diagnostic centers
—Birthing centers
—Hospices

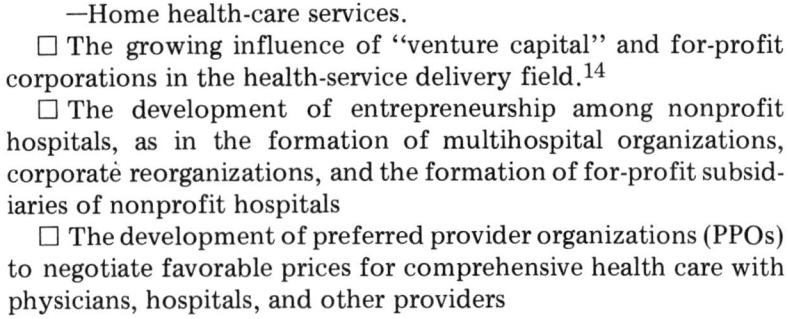

—Home health-care services.
☐ The growing influence of "venture capital" and for-profit corporations in the health-service delivery field.[14]
☐ The development of entrepreneurship among nonprofit hospitals, as in the formation of multihospital organizations, corporate reorganizations, and the formation of for-profit subsidiaries of nonprofit hospitals
☐ The development of preferred provider organizations (PPOs) to negotiate favorable prices for comprehensive health care with physicians, hospitals, and other providers
☐ The renegotiation of health-insurance benefit packages with cost-sharing incentives to promote responsible use of services and active management of benefits through programs such as preadmission certification.

The health industry in the past was characterized by numerous small and medium-sized organizations, mainly nonprofit in nature. The more than 7,000 hospitals that existed at that time were independently owned and controlled. Solo or small-group practice for physicians was the dominant practice pattern. Home health agencies were community-based individual agencies, usually visiting nurse services.

For-profit hospitals and hospital chains, and for-profit home health agencies existed before the mid-sixties, but they constituted a very small proportion of the health industry except for the drug, device, and medical supply sectors. The Medicare/Medicaid legislation stimulated the growth of for-profit hospitals, nursing homes, renal dialysis centers, and home health agencies. Three reimbursement factors encouraged the growth of for-profits: payment of interest on borrowing; payment for return on equity; and payment of depreciation on the facilities, which could be revalued on sale. In the last 20 years, chains and corporations have purchased facilities and services previously owned by individuals and small groups. With the increased incentives to move services out of hospitals, for-profit HMOs, surgicenters, and urgent-care centers have emerged.

The nonprofit hospital industry, which heretofore confined itself to the operation of single units and to hospital-based services, is now seeking to emulate the pattern of the for-profit sector. Multihospital organizations are forming; hospitals are reorganizing and forming for-profit subsidiaries to provide products and services; hospitals are opening free-standing urgent-care centers and surgicenters and, increasingly, are employing physicians. Both the profit and nonprofit sectors are offering services—management, financial, data, and purchasing—to other health providers.

The increased financial pressures of excess capacity and reimbursement will, over time, force closures, mergers, and growth of the hospital corporate umbrella, to compete in the new market environment.

With the growth in supply of physicians, the increasing debt burden for education, and the rising costs of practice, young physicians have been seeking and accepting salaried arrangements in hospitals, HMOs, multispecialty group practices, and urgent-care centers. The balance between the individual entrepreneur and the corporate entrepreneur in health care is shifting.

Increasingly, these institutions are actively competing to provide health services to health-benefits plans, to form preferred provider organizations, and to control price increases. Some are now underwriting health insurance.

The changes occurring in the structure of health-services delivery and the active management of health care by employers and unions are beginning to have a profound effect on utilization of services and costs. Hospital occupancy rates and lengths of stay are declining, day surgery is increasing, and cost increases are moderating. The actions taken under public programs, combined with the increasing intervention by the private sector, can accelerate these early positive trends.

The Need for Further Policy Efforts

The hopeful tone of the preceding paragraphs, and the proliferation of writings on health-care policy, may prompt the reader to ask, "Is there anything left to say?" Indeed, there is. First of all, no one knows how long-lasting the recent changes will be. Second, while hospital admissions and stays have declined, it is not clear that hospital costs have come down or will stay down. Third, pressures to expand hospital capacity and medical technology continue unabated.

On a positive note, there are still issues to discuss because health-care management has come to be viewed in a new light. The original impulse was to get health-care *costs* under control, and that has begun to happen. But in the process employers and unions have discovered that quality of care, based on intelligent choice among well-thought-out alternatives, is the best guarantor of cost-effective care.

Employers and unions now sense that they *can* make the system work better, at least as far as private patients are concerned. Previously, employers had neglected health-care management almost entirely, because they felt they lacked the requisite knowledge; that knowledge seemed to be proprietary to doctors, hospital leaders, insurers. Moreover, the providers and insurers made all the key decisions about diagnosis, treatment, and reimbursement. Finally, managing health care seemed irrelevant; health care was just part of the wages and benefits package.

Another major point: If health care is to be managed, it will require a substantial, continuing investment of money, people, and time—for example, collecting and analyzing data, mounting a prevention program, educating employees about the health plan. Many of our recommendations will similarly require investment. The payoffs will come, as a Pfizer Inc. spokesman has pointed out, in several forms: improved employee productivity and performance, improved employee health, reduced accidents and illnesses, improved employee relations, better community relations, receptivity to benefit changes, and reduced health-care costs.

Scope of the Report

Readers may grant that there is something new to say about health-care management but question the credentials of Work in America Institute to say it. The Institute, after all, has no visible expertise in health care.

In fact, when General Electric first urged the Institute to undertake a national policy study of health care, we raised the same doubts. We were assured (correctly, as events have shown) that there was no need for concern. The request came about because GE, as co-leader of The Business Roundtable Health Initiatives project, had concluded that it was highly desirable to involve additional employers in the work, to reach out to unions, and to explore in greater detail the findings of The Business Roundtable Task Force on National Health. Three other companies associated with The Business Roundtable Health Initiatives project—Atlantic Richfield, Metropolitan Life Insurance, and Pfizer Inc.—joined GE in supporting the study.

The point of the report is that it deals with the *management* of employee health care *in the workplace.* In that respect, the report falls directly within our competence. On matters of health care per se, we have relied on the advice of a few dozen people with outstanding qualifications in that field, whose names appear in the lists of National Advisory Committee members, contributors of background papers for the study, and Design Committee members at the beginning of the book.

As a first step, we convened a design committee of experts from industry, labor, and the health-care professions in December 1982 and asked whether a Work in America Institute policy study could contribute anything useful to a field so far outside of our normal sphere. The committee discouraged us from tackling the health-care "system" but strongly recommended a study of actions that employers and unions can take to improve health care in the workplace. They suggested four areas in need of attention: comprehensive corporate strategy for health care, incentives for better management of health care, alternative delivery systems, and joint employer-union (or employer-employee) efforts to improve health care.

The policy study and this report have stayed within those wise guidelines, despite subsequent urgings to wander, with one exception: In the course of the study we reached the conclusion that employers and unions, acting jointly and *locally*, could help to shape the overall system in addition to events within their own organizations. In particular, we believe that they can check the undisciplined growth of hospital capacity and of medical technology.

This is not to say that employers and unions should not also try to act at state and local levels. They have done so, and are doing so, with some success. But we recognize that certain major problems of health policy—for example, coverage for the uninsured, uncompensated care, cost shifting between public and private patients, coverage for people not in the work force, and making physicians allies in the reforms under way—are not within the power of employers and unions to solve.

Work in America Institute has always made it a point not to dwell on problems unless it is prepared to recommend solutions. The Institute's role is that of social technologist rather than social scientist—and so it is in this report.

Guidelines

The recommendations in this report are pragmatic in the best sense, but they are not ad hoc. They grow out of certain basic views about (1) relations between employers, employees, unions, and the community; (2) the value of personal choice in health care; and (3) the operation of the U.S. health-care "system," as follows:

1. The overriding objective of health-care management is not cost containment but cost effectiveness, that is, extracting from each dollar spent a higher quality of health and extension of life for all employees.

2. Health care, as currently delivered, is replete with unnecessary services and charges. Quality of care can be improved at the same time that costs are reduced.

3. Health care is the most expensive single item in the benefits package.

4. Wages and fringe benefits are paid out of a single pot. Funds applied to one benefit are diverted from others, or from wages.

5. Employees are capable of making intelligent decisions about health care—provided that they receive the requisite information, from sources they trust. Employers and unions should make every effort to enhance that capability and its expression.

6. It is now accepted that employees want to be involved in decisions affecting their work lives—and all the more so in decisions affecting their health and that of their dependents.

7. The most effective form of involvement is joint employer-union action (where employees are represented), or joint employer-employee action in other instances.

8. Because health care touches the lives of employees so intimately, efforts to change health-care benefits require employee cooperation before, during, and after the change. For the same reason, changes should allow for wide variation in individual preferences, customs, and practices.

9. Where health care is concerned, the word "employees" is often shorthand for "employees, their dependents, and retirees." The latter categories can account for as much as two-thirds of corporate health-care costs.

10. Since most people resist hospitalization unless it is absolutely necessary, employees should view reduced hospital utilization as a benefit, not a loss.

11. The key to permanent improvement in health-care management is more intelligent *behavior* on the part of *users*, which in

turn will compel hospitals and physicians to perform more cost-effectively.

Content of the Report

We believe that the content of the report adds up to a coherent approach to improving the management of health care in the workplace. Recommendations will be found at the end of each chapter.

Chapter 1, *Health-Care Management in the Workplace: An Overview*, recounts how the U.S. system of health care got out of control, summarizes current efforts to restore control and strengthen competition, and describes the focus of this report—what employers and unions can do to reform health care.

Chapter 2, *Joint Action: Employers/Employees/Unions* stresses that since employers and employees are inescapably partners in health care, they have more to gain from working on it as a team than at arm's length; that, when employees are unionized, employer-union partnership works best; and that the unions have demonstrated readiness for joint action. Indeed, some unions began pressing employers for action long before most employers became alarmed about health-care costs.

Chapter 3, *Designing Health-Care Benefits*, shows how hospital insurance inadvertently created the monster of overhospitalization and how the judicious use of alternative delivery systems and cost sharing can beat it back.

Chapter 4, *Managing Health-Care Benefits*, points out that having a good benefits plan is only a first step; the plan must be managed with the same kinds of tools as other plans, including timely, relevant information. Some common-sense methods are recommended.

Chapter 5, *Work-Site Wellness Programs: Fad or Future?* suggests ways of determining which kinds of wellness programs have the greatest value for each particular organization.

Chapter 6, *Health Maintenance Organizations*, recommends ways of hastening the spread of these useful institutions, but cautions against idealizing them. It suggests that some of their strengths, if applied to non-HMO practice, might heighten competition.

Chapter 7, *Corporate and Union Strategies for Health Care*, shows how some leading corporations, with encouragement from The Business Roundtable—and how some unions, acting with extraordinary foresight—have been developing an integrated approach to health care, in which the whole is greater than the sum of the parts.

Chapter 8, *Reducing Excess Hospital Capacity*, argues that hospital capacity nationwide is far in excess of needs, that this is a major factor in health-care costs, and that employer-union cooperation at the local level can resolve the problem without unduly harming hospital employees or the impoverished and underserved parts of the community.

Chapter 9, *Controlling the Use of Medical Technology*, deals with another major cause of high costs: the uncontrolled proliferation of medical technologies without regard to cost-effectiveness, sometimes even without regard to medical effectiveness. It is suggested that cooperative action by employers and unions at local level can help get the situation under control.

NOTES

1. Paul Starr, *The Social Transformation of American Medicine* (New York: Basic Books, 1982).
2. Michael Zubkoff, ed., *Health: A Victim or Cause of Inflation* (New York: Prodist, 1976).
3. National Academy of Science, Institute of Medicine, *Costs of Education in the Health Professions* (Washington, D.C.: National Academy of Sciences, 1974), chapter 1.
4. U.S. Department of Health and Human Services, Health Resources Administration, *Summary Report of the Graduate Medical Education National Advisory Committee*, DHHS (HRA) 81-651, September 1980.
5. "84th Annual Report on Medical Education in the United States, 1983-84," *Journal of the American Medical Association* 252 (September 28, 1984).
6. National Center for Health Statistics, *Health U.S., 1984*, DHHS Publication No. (PHS) 85-1232, tables, 72, 73, 78-81, 82.
7. Herman M. Somers and Anne R. Somers, *Medicare and the Hospitals* (Washington, D.C.: Brookings Institution, 1967).
8. Zubkoff, *Health: A Victim or Cause of Inflation*.
9. Social Security Amendments (1983).
10. Tax Equity and Fiscal Responsibility Act (1982).

11. Bruce C. Vladeck, "Medicare Payment by Diagnostic Related Groups," *Annals of Internal Medicine* 100 (April 1984).
12. Ibid.
13. Institute of Medicine, "Controls on Health Care," in *Papers of the Conference on Regulation in the Health Industry, January 7-9, 1974* (Washington, D.C.: National Academy of Sciences, 1975).
14. Arnold S. Relman, "The New Medical Industrial Complex," *New England Journal of Medicine* 303 (October 23, 1980).

2. JOINT ACTION: EMPLOYERS/ EMPLOYEES/UNIONS

The pattern of employer-union relations with respect to health-care costs is strikingly reminiscent of what happened in the case of productivity growth. In both cases, employers suddenly became aware of the problem and tended to blame employees, and especially unions. Gradually, as they studied the problem in more detail, they perceived that some of the responsibility may have been their own. Later still, it became clear that the roots of the problem lay in management's own omissions and commissions, that employees and unions had been reactors rather than actors, and that, to find solutions, management had to take the lead in partnership with employees and unions.

Unions have as direct an interest in health-care management—including cost containment—as employers. Union members are, after all, the recipients of the benefits the unions negotiate. Further, union members pay indirectly, since health costs are a major part of benefit costs. As health-care costs inflate, they displace other benefits or even wage payments. Coverage, quality of care, accessibility of care are of vital importance to the membership, all the more so when health care costs so much and employment security is so doubtful. The issue is particularly sensitive because health benefits provide protection for family health and survival against unpredictable risks and great hidden dangers. Individual employers and local unions clash on this issue when employers try to diminish the coverage or raise employees' share of the cost of existing coverage. At national, state, or community levels, however, union-management cooperation on cost containment remains strong.

Joint Action at National, State, and Community Levels

Early in 1985 the AFL-CIO established a special committee charged with developing health-care cost-containment techniques that place the burden of containment on providers and insurance companies rather than on patients and health-care employees. Chairing the committee is the president of the Service Employees International Union (SEIU), a union which has fought for pre-admission testing, second surgical opinions, concurrent review, alternative delivery systems, fee schedules, negotiated payments, hypertension screening and treatment, and drug and alcohol treatment programs.[1] These programs are also favored by employers.

Joint employer-union actions to improve the management of health care have been undertaken at national, state, community, company, and plant and office levels.

The best-known example on a national scale is the report of the Labor-Management Group organized by former Secretary of Labor John T. Dunlop in 1978 with top leaders of the AFL-CIO, United Auto Workers (UAW), Teamsters, and some of America's most prestigious corporations. After a nine-month study the group published the following recommendations:

☐ Support prospective reimbursement.
☐ Develop and/or remove barriers to health maintenance organizations (HMOs).
☐ Work at all levels to enhance the effectiveness of the health-planning system.
☐ Develop provisions for preadmission testing programs (and eliminate disincentives).
☐ Support pilot second-opinion programs.
☐ Develop provisions for utilization review and ensure that carriers and hospitals develop concurrent utilization review programs.
☐ Expand alternatives to inpatient hospital treatment.
☐ Support training programs for nurse practitioners and physician assistants; encourage education about cost advantages of using physician alternatives.
☐ Support appropriate use of medical technology (particularly on hospital costs).
☐ Reduce medical malpractice.
☐ Provide health education for employees.[2]

Growing out of the Dunlop group, a smaller Group of Six was formed in January 1982 with the following membership: AFL-CIO, American Hospital Association, American Medical

Association, The Business Roundtable, Blue Cross and Blue Shield Association, and the Health Insurance Association. Its mission is to provide a forum for leaders in the field of health care; its top priority is to foster the establishment of coalitions at state, regional, and local levels to control health-care costs. Currently, almost 150 coalitions are in operation; at least 45 of them have labor representation.

The Group of Six urges coalitions to:
☐ Redesign insurance plans to emphasize preventive, primary, and home care.
☐ Modify facilities and the number of hospital beds to make the most appropriate use of community health-care resources.
☐ Develop more effective programs of health promotion and disease prevention in the workplace.
☐ Improve access to care for the unemployed and others lacking adequate access.

Another national joint-action body, but within a single industry, is the Joint Labor-Management Committee of the Retail Food Industry Trust. Formed in 1974, the committee primarily collects data on wages and benefits and facilitates industry-wide collective bargaining. In 1978 the committee commissioned a study of ways to reduce health-care costs without cutting benefits. Since then, the committee has conveyed the survey findings through meetings of senior trustees and has tried to educate trustees, employees, and providers. In 1984 it carried out pilot studies to pave the way for local health-care action in Providence, Rhode Island, and in St. Louis.

A good example of a state organization is the Economic Alliance for Michigan, whose mission is to encourage more rapid expansion and diversification of the state's economy and to make the costs and conditions of doing business in the state more competitive. A former vice president of Ford Motor Company and a former vice president of UAW cochair the 90-member board of CEOs and union leaders. Action-oriented, the Alliance conducts legislative, educational, and promotional programs and is helping to improve the business climate in the state.

Health-care costs are a top priority of the Alliance, because employers in Michigan spent $4.7 billion in direct costs for health care in 1983. The organization selected three key objectives: joint business-labor promotion of enrollment in HMOs; passage of state legislation to permit the establishment of preferred provider organizations (PPOs) and to strengthen the state's ability to control capital spending on health care. With only 4 percent of the state's population enrolled in HMOs, the Alliance has held conferences around the state, collected data, and educated the public. It managed, after a long struggle with providers, to get PPO legislation passed. On capital spending, it lost out to the providers on the first round but is preparing to try again.

The most rapidly growing vehicle for joint employer-union action in health-care management is the coalition. A coalition is an organization of employers, with or without other directly interested groups, designed to concentrate pressure for change, usually on a county or community base. Of the more than 110 coalitions operating in 1985, at least 45 have unions participating. Those with union participation have the strongest financial backing, which indicates that the unions have supported their convictions with funds.[3]

Since most employer-union coalitions are of recent origin and tend to focus on long-range problems, one cannot yet expect major changes. However, some highly useful developments are under way:

☐ The Capital Area Coalition, in Washington, D.C., includes members from all interested parties: Blue Cross/Blue Shield, insurers, hospitals, physicians, unions, and others. One of its co-chairs is a top officer of the International Typographical Union. The coalition is developing, with support from the CIGNA Corporation, a guide to costs per diagnosis and per procedure in the hospitals of Washington and its suburbs. In June 1985 it held a day-long session to examine preadmission certification and other utilization controls.

☐ The Minnesota Coalition, based in Minneapolis, is a statewide organization which has been active mainly in the Twin Cities. Unions, employers, government agencies, providers, and insurers are members; the board includes representatives of UAW and the AFL-CIO. The coalition has testified before the state legislature in favor of competition; it has surveyed employers introducing changes into health-care plans; and it has surveyed employees about what they like and dislike in their plans. Unions have been particularly active in committees dealing with health education, a community buyer system, and health costs in the school system.

☐ The Kansas Employers Coalition on Health, based in Topeka, has no unions as formal members, but has cooperated with building-trades unions and the machinists. (A union-only consortium covering several cities in Kansas sought to negotiate PPOs with providers, since they were not inhibited by the state statutes which block employers and insurers from doing so.) The coalition is making progress on state legislation to allow PPOs. It is gathering cost data directly and voluntarily from providers, on the basis of diagnostic related groups (DRGs), for all under-65 employees; the data have been given voluntarily because Blue Cross/Blue Shield began paying by DRGs early in 1984.

☐ The Iowa Business-Labor Coalition on Health, whose members include the UAW and the AFL-CIO, was formed in 1981 to foster statewide activities for health-care cost management. It guides eight local coalitions in major population centers.

☐ Other notable coalitions with active union involvement are in Cleveland and Atlanta.

Joint Action at the Level of the Company or Site

The path to joint employer-union action on health-care management in the company or work site is obstructed by vehement differences of opinion on cost sharing. Employers whose plans provide first-dollar coverage have demanded that employees pay deductibles and coinsurance; those with cosharing plans have demanded higher contributions by employees. Unions such as the United Auto Workers, the United Steelworkers of America (USA), and the Communications Workers of America (CWA) have stood their ground and refused to give up first-dollar coverage; they have offered, as an alternative, to enter into, or to expand, joint employer-union programs to contain costs.

Some employers have not pressed for cosharing, seeing it as an irritant, and have themselves offered joint action. Many have imposed cosharing on nonunion employees and fought off union objections in collective bargaining. Others have softened the blow by matching cosharing with arrangements for partial or total reimbursement at the end of each year.

Employers see cosharing as a proven, effective, and fast method of reducing corporate health-care costs. Health care has become the fastest-rising and the largest single cost in the benefits package, and thus the most attractive target for cost reduction. Cost sharing not only reduces the employer's costs immediately; it also reduces them on a continuous basis by causing utilization to decline. Moreover, it does not affect the health of most employees, so the employer feels no guilt. Finally, since 80 percent of the work force is nonunionized, cost sharing can be introduced quickly and unilaterally.

Cost sharing is so tempting, unfortunately, that few employers can resist it, although it may foreclose joint employer-employee endeavors, noted later in this chapter, which would probably be worth a great deal more to the employer in the longer term.

Unions grant that costs have gotten out of hand, but they deny responsibility for it. They raise the following objections:

1. Requiring cost sharing where there was none before, or increasing the amount of cost sharing, represents "takebacks." If imposed during the life of a collective agreement, they unfairly alter the agreement. Even if health-care benefits are paid by the employer alone, they are not gratuities; they were traded in exchange for forgoing other economic gains, such as wages and benefits. Nor is a dollar's worth of health-care benefits equivalent to a dollar's worth of wages; health benefits are tax free, whereas wages are taxable. So a dollar of cost sharing is equivalent to approximately $1.25 of wages. Moreover, health benefits are probably the most highly valued gains a union can boast. Therefore, when management prevails and introduces cost sharing, the union is apt to lose face.

2. Employees don't have to be induced to spend health-care dollars more prudently. They don't go to the doctor because they like to; they go when they have to. From that point on, the doctor makes all the decisions. The doctor decides what treatment is necessary, whether the patient should be hospitalized and for how long, what tests must be administered and where, and what drugs, if any, must be taken. If the employer wants more prudent spending, it should direct its attention to the doctor and the hospital and induce *them* to behave more prudently.

3. It is unfair and untrue to imply, as the argument about "spending more prudently" implies, that unions are indifferent to health-care costs. The record shows that some unions, such as the United Auto Workers and the International Ladies' Garment Workers' Union (ILGWU), have been more cost-conscious than employers. The UAW advocated and helped to establish HMOs a generation ago, while employers turned a deaf ear. The ILGWU, whose benefit funds are much more meager than those of unions

in the heavy industries, has conducted a thorough but highly cost-effective program of care for many years. The Service Employees International Union has taken similar steps in recent years. Many unions have urged and joined in programs of utilization control, including preadmission certification, mandatory second opinions, generic drug prescriptions, and so on.

Interpreting the Rand Experiment

In the running dispute between employers and unions over cost sharing of health care, employers frequently cite the findings of the Rand experiment to bolster their arguments for cost sharing, while unions feel compelled to attack the Rand experiment on structural and procedural grounds because the findings seem adverse to their position. The experiment, conducted by the Rand Corporation between 1974 and 1981 under a grant from the U.S. Department of Health and Human Services, measured the consumption of services by 7,703 persons in 2,757 families enrolled in health-care plans at six geographically dispersed and varied sites. Most of the people were enrolled in fee-for-service insurance plans, with varied levels of cost sharing; some, in a fully prepaid HMO. In our opinion, both employers and unions may be misinterpreting the Rand experiment findings.

The shared misinterpretation may stem from the fact that Rand reported its findings on service consumption in two separate articles, spaced two years apart. The first article dealt solely with health-care consumption among families enrolled in several fee-for-service plans.[4] Each plan required a different degree of cost sharing, ranging from zero to 95 percent. The second article compared the health-care consumption of families in the fee-for-service plan requiring zero cost sharing, against the health-care consumption of families enrolled in a prepaid group HMO, also with zero cost sharing.[5]

What did Rand actually find?

The first article showed that, as among families in the various fee-for-service plans, health-care consumption was closely correlated with the degree of cost sharing: the lower the degree of cost sharing, the greater the consumption of health care. In particular, families with zero cost sharing consumed about 50 percent more than families on the least generous plan. This is the conclusion cited by most observers in the debate on cost sharing.

No evidence has yet been adduced to challenge the finding

that higher degrees of cost sharing in fee-for-service plans are correlated with reduced consumption of health care. But stating the correlation does not answer the more urgent question: What *accounts for* the correlation?

The answer comes in the second article, which found that the HMO families, despite zero cost sharing, consumed 28 percent less health care than the families in the fee-for-service plan with zero cost sharing. In other words, two different types of plans, both having zero cost sharing, produced vastly different rates of health-care consumption. It follows that cost sharing, in and of itself, does not account for the wide variations in health-care consumption from plan to plan.

If cost sharing per se is not the explanation, what is? According to Rand, "The principal cause of the difference (between HMO and fee-for-service consumption) was in the hospital admission rate, which was 40 percent less in the HMO group than in the fee-for-service (zero cost sharing) group."

It turns out that different rates of hospital admission are also the principal reason for the different rates of health-care consumption among the fee-for-service plans. Thirteen percent of families on the zero cost-sharing plan were hospitalized, whereas only 8 percent to 10 percent of those on the less generous plans were hospitalized. This means that hospital admissions on the zero-cost-sharing plan were 30 percent to 50 percent higher than on the less generous plans. The higher the degree of cost sharing, the less often the families visited their physicians; the less often they visited their physician, the less they were hospitalized. Interestingly, HMO families visited their physicians more often than their counterparts on fee-for-service plans. Rand does not explain this discrepancy, but it is a fair guess that some of the HMO office visits were used either for prevention or procedures that fee-for-service physicians would have had performed in the hospital.

What accounts for the different rates of hospitalization between HMOs and fee-for-service plans? The answer, according to Rand, is "a different style of medicine." The difference in style amounts to the fact that HMO physicians make much greater efforts than fee-for-service physicians to find alternatives preferable to hospitalization (see chapter 6, "Health Maintenance Organizations").

It follows that employees (and their dependents and retirees) do not need to be persuaded to act prudently in matters that literally affect their lives; they need to be shown how. They lack independent means of knowing which providers are cost-efficient and of high quality. They are fearful of challenging a provider. They do not know which tests or surgical procedures can safely be performed out of the hospital. They often are uninformed about how to determine whether an illness does or does not require medical care. Millions of otherwise intelligent people do not understand how "life-style" affects health. And beyond information, they may need inducements and reinforcements to help them change to better self-management.

Thus, the Rand findings can be used to support both sides of the cost-sharing argument. The unions can legitimately assert that it is physicians, not patients, who determine how much health care is consumed, and that such mechanisms as HMOs, preadmission certification, and second opinions produce greater economies than cost sharing per se. Employers can legitimately claim that cost sharing reduces health-care consumption in fee-for-service plans by reducing hospital utilization.

But cost sharing is a blunt instrument. Employers can achieve the same or better results, without the drawbacks, by using HMOs and other utilization controls and by creating financial incentives for best practice, or disincentives for the opposite. The drawbacks are those mentioned earlier: the souring of employer-union relations and the loss of tax advantages. Cost sharing unnecessarily impedes joint employer-union action, action which has much greater potential for both short- and long-term cost saving than any unilateral action can have. Unilaterally imposed cost sharing also undercuts the efforts of employers who seek a participative approach to employer-union relations in general.

It was pointed out earlier that requiring employees to share (or to increase their share) in the cost of health care takes more out of their pockets than meets the eye, because benefits are tax-deductible to the employer but not to the employee. The employer also may be harmed, because tax-deductible fringe benefits help to attract new employees and cost sharing may reduce the value of the compensation package more than the employer saves up front.

Models of Employer-Union Joint Action

Efforts by an employer to alter employees' behavior in such a sensitive area are bound to encounter mistrust and resistance.

The normal reaction is to assume that the employer is trying to save money at the employee's risk. The best way to persuade employees that a change in the benefit plan works to their advantage as well as the employer's is to bring them into the process of designing and executing the change—preferably through their union, if there is one. Where there is no union, a committee structure of some kind can be substituted.

General Motors/United Auto Workers. Probably the most comprehensive program of intracompany joint action is that established by contract between General Motors and the UAW.[6] The agreement of September 1984, which was negotiated while this study was in progress, rejects cost sharing but adopts the goal of containing health-care costs "by changing the wasteful and inefficient way health care is delivered." It sets "a target of 10 percent reduction in costs and holding increases to an adjusted inflation factor."

☐ The agreement includes an Informed Choice Plan, under which UAW members will be able to choose either the traditional fee-for-service plan, an HMO, or (after January 1, 1985) a PPO. The scope of the traditional plan, providing first-dollar coverage, is expanded, but now requires "predetermination of all non-emergency and non-maternity hospital admissions and certain other services," as well as "concurrent review of the length of stay in the hospital."

☐ PPO coverage is somewhat broader than that of the traditional plan. "The PPO assumes responsibility for conducting utilization reviews, predetermination of certain services, and monitoring for quality care."

☐ Several joint committees are charged with responsibility for:
  —Adding new items to, or subtracting them from, the traditional plan's lists of medical equipment and procedures.
  —Devising a special incentive program to provide additional maternity benefits and encourage shorter maternity stays in hospital.
  —Developing a voluntary case-management program to identify potential high-cost medical cases and explore alternative treatment plans to provide care in a more cost-effective manner while maintaining or improving the quality of care.
  —Developing procedures to monitor the quality of HMO

performance and ensure that GM/UAW members receive quality care.

—Developing and implementing an educational program for the purpose of encouraging the use of health-care coverage in an informed and cost-effective manner; defining the options available under the Informed Choice Plan; and developing a wellness or health-promotion program (initially on a pilot basis).

—Addressing problems of patient delays and extra billing in the Retirees Service Program.

To ensure that the joint committees are empowered to get things done, a jointly administered fund of $1.5 million was established in 1982. Money from that fund has already purchased a cost comparison of various health-care programs; a comparison of dental capitation and fee-for-service plans; an examination of the financing and quality of care in HMOs; a management information system to analyze medical and hospital costs and utilization; a communications program to inform employees about the informed choice program; joint training of benefit-plan administrators; and pilot HMOs for retirees.

Under previous agreements a joint program has dramatically reduced the rate of sickness absenteeism. Any GM/UAW member whose absence continues longer than expected is required to be examined by a jointly selected medical clinic. The clinic's word is final. Other pre-1984 joint actions include:

☐ Preadmission testing
☐ Concurrent utilization review
☐ Mandatory second opinions for surgery
☐ Expansion of dental HMOs and PPOs
☐ Preauthorization at all hospitals in a three-county area near Flint, Michigan
☐ A pilot program under which incentives were paid to favor ambulatory rather than inpatient surgery for 26 common procedures

American Telephone and Telegraph (AT&T)/Communications Workers of America (CWA)/International Brotherhood of Electrical Workers (IBEW). A precipitating factor in the AT&T strike of 1983 was the union's refusal to accept management's demand for cost sharing in the health plan. Nevertheless, the eventual agreement established a joint AT&T/CWA/IBEW steering committee on health-care cost containment and a similar working committee at each Bell company to examine the major causes of health-care costs in the company, recommending containment measures, participating in local coalitions and other action groups, cooperating in community-based utilization review programs and in PPOs, and encouraging the use of HMOs.[7]

In 1984 the steering committee chose four areas of concentration: data analysis to pinpoint the causes of health-care cost, review and testing of plan design features, educational programs, and community initiatives. Particular measures under consideration include:

- Encouragement of employee participation in HMOs that meet standards
- Allowing employees to audit their hospital bills and share in savings
- Reducing costs of administering the plan
- Reviewing personnel practices that bear on health costs, such as disability and absenteeism
- Preauthorization of hospital admissions and concurrent review
- Alternative delivery systems
- Incentives for second opinions and outpatient treatment
- Cost sharing
- Educational programs for health promotion
- Counseling employees who have to make medical-treatment decisions
- Advising employees how to make best use of the plan

Big Steel/United Steelworkers of America (USA). In the 1982 negotiations with Big Steel, the United Steelworkers of America, although reeling from the recession, rejected givebacks in health benefits. But it agreed to establish joint labor-management health committees at each company, which were authorized to put into effect appropriate cost-containment measures. Among the measures to be investigated were: second opinions on surgery, preadmission screening, and utilization review.

In other instances unions agreed to cost sharing at the same time that they entered into joint employer-union programs for health-care management.[8]

Teamsters/Master Freight, 1982. Under the agreement, part of the cost-of-living allowance (COLA) increase was diverted to pay for increased employer contributions to pension, health, and welfare plans; if required contributions exceed the COLA, employers will make up the difference. In order to minimize the diversion from COLA, local joint funds adopted such measures as:

☐ Deductibles and cosharing for hospital patients who fail to obtain preadmission testing
☐ Coordination of benefits with state no-fault auto insurance policies
☐ Negotiation of sizable discounts for early payment of hospital bills
☐ Audits for all hospital bills, and efforts to negotiate reductions of bills exceeding $25,000
☐ Full coverage of costs of a second opinion before surgery
☐ Reserve power to make second opinions mandatory for certain nonemergency procedures [9]

Oregon Public Employees Union (Local 503, Service Employees International Union [SEIU])/State of Oregon, 1980. Under the agreement, covering 17,000 employees, deductibles and cosharing have been increased, but the following joint efforts are in practice:

☐ Mandatory screening for all nonemergency hospital admissions by an arm of the local medical society. If screening raises major questions, a fully paid second (or even a third) opinion is required.
☐ All hospitalizations must be reviewed and authorized before admission by the state's insurer. The insurer also predetermines length of stay.
☐ Programs for disease prevention, testing, education, exercise, stress, hypertension, and encouragement of home health services.

As a consequence, many conditions that used to result in short-stay hospitalizations are now treated on an outpatient basis—a reduction of unnecessary surgery. The annual rise in the overall cost of the plan has been brought down from 41 percent in 1981 to 27 percent in 1982 and 6.5 percent in 1983.[10]

Aluminum Company of America (ALCOA)/Aluminum, Brick, and Glass Workers International Union (ABGWIU), 1984. This agreement creates copayments and deductibles, offset by a reimbursement fund of $700 per annum per employee. The fund can be applied to cover not only the copayments and deductibles, but also cosmetic surgery, orthodontics, and other previously uncovered costs. It is expected that almost 90 percent of covered employees will recover unused balances, which are returned as taxable cash to the employee or to a special IRA. The parties have operated successful joint committees on safety since 1976 and on disability management since 1983.[11]

Procedural Aspects of Joint Action

Employers and unions that join forces to improve the management of health care should appreciate that designating a joint committee is only the first step. To function well, committees need financial resources, some degree of training, and sound business procedures.

Not every committee can hope for the large sums available, for example, to the General Motors/UAW committees, but the GM experience shows the value of being able to buy expert consultation. Managers and union representatives cannot rely on their own experience alone to devise workable solutions to health-care problems. Whether they choose to examine hospital utilization, hospital-management practices, or providers' performance statistics, they must fall back on expert guidance.

In addition to knowledge about health care, the committee members require skill in working as a semiautonomous group. Each committee is charged with a major responsibility, which it must manage over a long period. The greater the ability of the group to work as a team, the better use it will make of the resources at hand. The group can also benefit from training in the arts of teamwork and problem solving as the need arises.

Finally, committees need to discipline themselves with basic managerial procedures, such as formal agendas, recording of decisions, follow-up of delegated responsibilities, regular reporting to the sources of authority, and the like.

The foregoing measures will strike a chord with readers who are familiar with employee involvement or quality-of-work-life programs and for good reason: joint committees to improve health-care management make a logical addition—or introduction—to employee involvement. They offer an opportunity for working together on a win-win basis, with the potential for large, tangible gains within a finite period of time.

RECOMMENDATIONS

1 Employers and unions should manage health care jointly, wherever feasible. Employees have as strong an interest as their employers in using health-care benefits cost-effectively, and the two parties can gain that end more successfully through joint action than separately. In particular, employee involvement in designing the health-benefit package leads to larger, longer-lasting cost reductions.

2 Cost sharing, although it has serious side-effects, should be thoroughly explored by the parties as an option which demonstrably lowers costs. Pros and cons should be examined. However, both sides should avoid taking rigid positions on this volatile issue, since it could get in the way of more valuable long-range improvements in the health-care program that require joint action.

3 Employers and unions should ensure that joint actions are supported with appropriate funds, training, and management methods. Investments in joint-action programs produce high returns and have lasting value.

NOTES

1. "AFL-CIO Healthcare Committee Announces Plans for Health Cost Containment Efforts," *Daily Labor Report*, May 24, 1985, pp. A-12-A-14.
2. William E. Hembree, "Joint Labor and Management Efforts to Control Employee Health Care Costs," an unpublished paper commissioned by Work in America Institute, May 9, 1984, p. 3.
3. Ibid., p. 4.
4. Joseph P. Newhouse et al., "Some Interim Results from a Controlled Trial of Cost Sharing in Health Insurance," *New England Journal of Medicine* 305 (December 17, 1981): 1501-1507.

5. Robert H. Brook et al., "Does Free Care Improve Adults' Health? Results from a Randomized Controlled Trial," *New England Journal of Medicine* 309 (December 1983): 1426-1434.
6. Hembree, "Joint Labor and Management Efforts to Control Employee Health Care Costs," p. 6.
7. Ibid., p. 8.
8. Ibid., p. 10.
9. Ibid., p. 11.
10. Ibid., p. 13.
11. Ibid., p. 8.

3. DESIGNING HEALTH-CARE BENEFITS

After five decades in which health benefits through private health insurance and self-insurance have grown in scope and depth, major changes are occurring in the underwriting and management of health-benefits plans. These changes are in response to the rapid ongoing inflation in health-care costs. Employers and unions, once passive insurers and payers of health care, are now actively seeking to manage health-benefits programs through several mechanisms.

☐ Design of the scope of benefits provided
☐ Incentives and disincentives to influence use of certain types of services, including changes in cost-sharing provisions
☐ Administrative control of certain types of services, such as hospital admissions
☐ Management of payments to providers

The Midwest Business Group on Health; General Motors and United Auto Workers; American Telephone and Telegraph, the Communications Workers of America, and the International Brotherhood of Electrical Workers; and other major employer and union groups have addressed benefits design and change comprehensively. Some groups, like the Service Employees International Union and the Teamsters, have pioneered specific programs, such as second surgical opinions and preadmission testing; others, like Bank of America, offer financial incentives to control utilization of services.

This chapter will describe changes occurring in the design of health-benefits programs, including alternatives to hospital care and financial incentives and disincentives. The succeeding chapters will describe management tools and provider relationships.

The Evolution of Health Benefits and Health Insurance

Until early in this century, and indeed until the 1920s, most medical-care services were provided in patients' homes and in physicians' offices. Hospitals were institutions for the dying or the poor and were avoided by most patients. While some health insurance was provided, mainly through fraternal organizations and some unions and employers such as the railroads, such coverage was rare.

The development of anesthesia, the advances made in surgery, and the revolution in medical education that followed the Flexner report of 1910 stimulated the development of the hospital as the central focus of medical care. A new generation of physicians trained in the Flexnerian mode saw the hospital as a laboratory for innovation in clinical medicine. However, the cost of care in hospitals was a barrier for many. In the late 1920s, consumers and hospitals saw the need for protection against high costs, and Blue Cross was formed by the hospitals and the public to spread the cost of hospitalization across a broader base of patients and communities through prepayment for care.

The technological advances made after World War II further spurred the development of hospitals. Of equal importance were two other factors: the Hill-Burton program initiated by Congress in 1948 and the rapid growth of health insurance. The Hill-Burton program was designed for two purposes: to assure that all communities in the United States had access to hospital care and to attract physicians and physician specialists to all communities.

Parallel Paths: Health Insurance and Hospital Utilization

The way health insurance developed has played a major role in health-care costs, health-care delivery, and utilization. Initially, health insurance, most notably Blue Cross, was designed to provide community-based insurance for "catastrophic costs"—hospital costs. Hospital costs in relation to the standard of living 40 or 50 years ago were high, and a hospital stay was a major burden for the family. At first, health-insurance policies provided hospital coverage only, and later these policies began to provide coverage for surgical services and anesthesia associated with surgery. The progression continued with coverage for hospital visits by physicians and coverage for in-hospital laboratory tests and X-rays. The growth and expansion of health insurance was spurred by the wage freeze during World War II. Unions negotiated for fringe benefits in lieu of wages, with health insurance a major demand. However, it was only during the fifties that payment for out-of-hospital services and major medical policies became more widespread.

Exceptions to this pattern were the few prepaid group-practice plans (now called health maintenance organizations, or HMOs) that existed before the 1960s. Some major comprehensive union-management contracts, such as those in the auto industry, covered a broad array of out-of-hospital services, and some union plans, like that of the International Ladies' Garment Workers' Union (ILGWU), had panel physicians who provided services at fixed fees or in health centers. In fact, until the very last moment of the Medicare debate in Congress in 1965, Medicare did not include physicians' and other medical benefits but was a hospital insurance plan.

Health insurance, as it developed in the United States, offered greater incentives to provide services in hospitals than in physicians' offices. Since the patient was reimbursed by insurance if an X-ray or other diagnostic test was performed in the hospital rather than in the physician's office, then, for the benefit of the patient, the service was provided in the hospital.

The hospital as a base for services also provided many advantages for the physician. The hospital provided the space, equipment, and ancillary personnel that could increase the productivity of the physician with no cost to the physician. Furthermore, even today, with rapid changes occurring, physicians are frequently paid higher fees for services provided in hospitals than for services provided outside of hospitals.

Hospitals also offered advantages to patients, namely quality control, although it was unevenly applied. The Joint Commission on Accreditation of Hospitals has been in existence for many years and sets minimum standards for fire safety, laboratories, minimum staffing, tissue committees, and so on. Many hospitals have mortality-review and peer-review committees that can deny or remove privileges from physicians whose practices are below standard.

Physicians in the United States, once licensed, are free to perform any service or procedure except for limitations imposed by hospital privileges. Unlike the situation in most European countries, specialists are not mainly hospital based, and primary-care physicians may provide specialty care. Specialists in the United States also practice primary care. Physicians practicing in private-office settings, particularly in solo practice or small groups, have few constraints on what they can do aside from their own caution or malpractice suits, and there is no routine peer review of practice except in some large group practices or HMOs.

The Paths Diverge

As technology and health-insurance benefits influenced the development of hospitals and the flow of patients to hospitals, so too are technological development and the changes in health-insurance benefits stimulating the movement of services out of hospitals. New anesthesia and certain surgical techniques developed in the last two decades now make out-of-hospital surgery feasible for many procedures. Lasers, CT scans, other noninvasive tests (such as diagnostic imaging of all kinds), and new chemotherapy developments make outpatient diagnoses and treatment possible. Certain other services thought to be substitutes for higher-cost hospitalization have been spurred by the development of insurance benefits. Medicare, for example, expanded home health services. Just as insurance and reimbursement policies perhaps unwittingly encouraged hospital use in the past, they are now discouraging hospital use and providing incentives for the "unbundling of services."

Although many of the so-called newer forms of ambulatory services are being hailed as methods of containing costs and expenditures, they are not new; they are merely a variation on services that used to be performed in individual physicians' offices or outpatient departments of hospitals, or provided by families or nonhealth personnel. For example, minor surgery, such as biopsies or removal of polyps, now performed in surgicenters, was done and sometimes still is done in physicians' offices; such surgery moved out of the office because reimbursement policies encouraged in-hospital surgery, either by requiring hospital stays for payment or by imposing cost sharing for outpatient surgery. Similarly, diagnostic testing was usually performed in the physician's office until insurance plans began to pay higher rates for having it done in the hospital.

The evolution and management of health-insurance benefits not only affected the use of different services but provided incentives to continually increase utilization. Cost or charge reimbursement to hospitals and fee-for-service payments to physicians also encouraged the increased use of services.

Rising benefit costs began to attract the attention of industry and labor in the late 1970s as premium costs began to escalate rapidly. Until that time pressures had consistently been toward the expansion of the scope of health insurance to cover outpatient medical services, dental services, and mental-health benefits, and toward the reduction of cost sharing by employees and dependents.

The population covered by health insurance also steadily increased until recently. First, large employer-union groups covered employees and then included dependents. Finally, coverage was extended to smaller groups. Until the 1950s most insurance was purchased from the nonprofit Blue Cross/Blue Shield plans, which can sell only health insurance, and are organized on a state or local basis. Commercial insurers, such as Prudential, CIGNA, and Metropolitan, who sell other types of insurance, such as life and disability, entered the field about the time of World War II.

Initially, Blue Cross/Blue Shield community rated health-insurance premiums. Critics claim that community rating—that is, averaging the risks within a geographic region—offers no incentives for utilization control. (For example, Medicare rates the entire population over 65 as a single community.) Commercial insurers began to compete actively with Blue Cross/Blue Shield by offering "experience-rated" health insurance to employer-union groups. Experience rating considers the risk of the specific employment group rather than the total risk of a geographic region. To remain competitive, most Blue Cross/Blue Shield plans shifted to experience rating. Some critics believe that experience rating encourages skimming of lower-risk groups. Under experience rating, large groups tend to have lower premiums per employee than small groups; low-risk white-collar industry has lower premiums than blue-collar industry; and younger work groups have lower premiums than older work groups.

A small number of employers and industry-union health-benefits plans have always self-insured. In recent years, however, there has been a rapid increase in the number of self-insured companies, for a number of reasons. Commercial insurance plans are subject to state taxes and, in some states, minimum benefit requirements. Self-insured plans are not subject to these factors. In addition, employers and unions concerned with "management" of health benefits, not just claims payments, found until very recently that large insurers were reluctant to make changes that the companies and unions deemed necessary to actively manage benefits and control expenditures. Today, some insurers have changed so radically that they merely carry out the administrative functions of health-care management for employers who self-insure.

Major changes are now occurring both in benefits design and in the active management of benefits. Most employers, unions, and insurance carriers could be classified as passive until very recently, paying the premiums and bills with little concern over utilization or price. The change to active management, particularly among large companies and unions, occurred with the pressures of the inflation and recession of the late 1970s and early 1980s and, more important, the increased competitiveness of the health industry. The new competitiveness is characterized by the expanded supply of physicians and facilities and alternative delivery systems, most notably the development of health maintenance organizations (HMOs) and preferred provider organizations (PPOs).

Benefit design requires a series of decisions, discussed in this and succeeding chapters. This chapter focuses on:
- Scope of services to be covered, including types of services
- Cost sharing for beneficiaries
    —deductibles
    —coinsurance
    —copayments
    —premium share
- Issues to be considered in design and redesign of benefits.

Scope and Types of Nonhospital Services

Very rapid changes are occurring in the types and sites of health care, spurred by three factors:
   —The increased supply of physicians and growing competition among physicians and between physicians and hospitals
   —Changes in reimbursement that are stimulating a shift from institutional to outpatient care
   —The entry of venture capital and the increase in proprietary interests

The discussion that follows will describe the different forms of out-of-hospital services, including the numerous types of ambulatory care centers, nursing homes, hospices, birthing centers, and home health services. Issues will be raised on cost, quality, and utilization. For many of these types of services there is little available data on costs, utilization, numbers of facilities, and quality. However, many of these services emerged quite recently, and data collection usually lags behind new developments.

Ambulatory-Care Settings

A variety of different types of ambulatory-care centers is developing. They include HMOs, which have existed for many years (see chapter 4, "Managing Health-Care Benefits"); surgicenters, which began under that nomenclature in the 1960s; birthing centers; free-standing dialysis centers, which were spurred by the 1972 Social Security amendments; and, most recently, emergency-care centers, urgent-care centers, and free-standing diagnostic centers.

HMOs differ from other ambulatory-care centers in two ways. They provide comprehensive ambulatory and inpatient care for their beneficiaries, and payment is on a prepaid capitation basis. The other types of ambulatory-care centers provide a specific service or a limited scope of services, and payment is on a traditional fee-for-service basis.

As was mentioned earlier, ambulatory-care surgery is not a new phenomenon. Its revival began in the late 1960s with the advancement of anesthetics. Hospitals were slow to establish day surgery, and their accounting methods spread indirect costs in a manner that made charges for ambulatory surgery in hospitals higher than in free-standing facilities.

Surgicenters were developed first by physicians. Surgicenters have quality standards just as hospitals do. Data on costs and quality indicate that surgery performed in these centers is safe and of comparable quality to in-hospital surgery and that the total cost per procedure is lower. However, the physician's charge portion is comparable for in-hospital or ambulatory surgery.

Until recently surgicenters were owned mainly by physicians and were in direct competition with hospitals. Recently, hospitals and investor-owned companies have been expanding their interest in surgicenters. Utilization problems could emerge as these centers spread. As studies by Dr. John Wennberg of Dartmouth Medical School and others have shown, elective surgery rates vary widely among similar populations. Unless second surgical opinions and rigorous peer review are extended to out-of-hospital surgery, the total volume and costs of surgery could increase as new centers spread. As routine surgery moves out of the hospital, the unit costs of in-hospital surgery could increase.

Emergency centers are for critically ill patients and must have close relationships to a hospital. These centers are not yet common in the United States.

Urgent-care centers, a very new phenomenon, provide a substitute for nonemergency care in hospital emergency rooms and for care in physicians' offices. The majority of centers that currently exist provide routine episodic care, although many provide laboratory and X-ray services. The centers are frequently staffed by part-time physicians. A number of these centers have sprung up in shopping centers.

While many hospital emergency rooms have developed relationships with practicing physicians for the care of patients after hours, with reports on findings sent to the primary physician, urgent-care centers do not yet have these linkages. Continuity of care for the patient and the interchange of information on preexisting conditions and on drug reactions are less likely with care provided in urgent-care centers than in physicians' offices or emergency rooms.

Physicians own many of these facilities. However, for-profit and not-for-profit hospitals are now starting these centers, as are health corporations. For hospitals and medical schools, satellite centers can reduce the inappropriate use of emergency rooms while retaining a certain amount of continuity of care, particularly if they serve as primary-care centers; the centers can also enable the hospital and/or medical school to be competitive with private physicians. However, as less costly services are moved out of hospitals, the cost-averaging potential is reduced and the costs for the remaining hospital outpatient services will rise.

Since urgent-care centers are a very recent development, there are no data on the number of these centers, volume, kinds of conditions treated, or costs. It is estimated that the potential dollar volume for urgent-care centers could range from two- to five-billion dollars a year.

The availability of physicians on evenings and weekends and the absence of the long waits and discomforts that characterize most hospital emergency rooms could stimulate the use of these

42   *Improving Health-Care Management in the Workplace*

centers for minor or self-limiting conditions and bring an increase in total utilization, costs, and charges to patients.

Technological advances have played a major role in the development of *free-standing diagnostic centers*. However, freedom from the rigorous regulation required in hospitals, related to both quality and certificate of need, has also stimulated the development of these centers. Services provided are often disease- or technology-specific, such as diagnostic imaging, nuclear magnetic resonance, cardiac catheterization. Diagnostic centers are not a new phenomenon—free-standing clinical laboratories and radiological centers have existed for years.

The proliferation of out-of-hospital diagnostic centers may increase the total volume of diagnostic services by making such services more attractive to both patients and providers: to patients because greater accessibility means less waiting time; to providers because they are less subject to legal regulations and budgetary stringencies than hospitals are.

Most of the services moving out of hospitals are "money makers," and once out of the hospital they are not subject to hospital reimbursement limits. Little is known about the total cost to society of diagnostic centers or the quality of care offered there. While there is routine data collection of utilization, cost, and morbidity for hospital-based services, data have not been collected on these centers.

Renal dialysis centers are unique but could be a precursor of the development of other types of outpatient treatment centers that provide one treatment modality, for example, plasmapheresis or chemotherapy. They are also unique because there is one dominant payer of services. Because Medicare is the dominant payer, control over price has been feasible. Medicare payment policies for renal dialysis have encouraged the development of free-standing dialysis centers and, until recent changes in Medicare, the payment process discouraged home dialysis, which is a less expensive form of care.

Important lessons can be derived from the Medicare experience with dialysis. These include:

- ☐ Rapid increase in utilization of dialysis once funding was assured, even in cases where social and ethical questions could be raised regarding efficacy of treatment for specific patients
- ☐ Stimulation of for-profit centers as a result of the level of

payment and provisions allowing for the return on equity capital
☐ Influence of payment levels on the siting of care

Considerable debate centers around the issue of whether *home health services* are substitute or additive services. A number of studies have addressed this issue, with conflicting results. Prior to Medicare, few third-party payers reimbursed for the costs of home health services. Medicare sought to reduce the need for high-cost hospitalization by paying for the alternatives of extended-care facilities (nursing homes) and home health services. Home health services before Medicare were provided through physicians' home visits, which had begun to decline after World War II; visiting nurse services; and families.

Over time, the home health-care services under Medicare have been liberalized, and the use of, and expenditures for, these services have risen rapidly. The new reimbursement method for hospitals is expected to increase the demand for, and use of, home health services. Because of the increased utilization and cost, the Reagan Administration and the Senate Budget Committee have proposed copayments under Medicare.

Recently, there has been an emergence of technology-specific services in home health care, starting with an increase in home dialysis, and now services such as total parenteral nutrition, cancer chemotherapy, and inhalation therapy. (The latter service has come under increasing criticism for excessive use in hospitals.)

Other Alternatives to Hospitals

Nursing homes are not a new phenomenon, although Medicare and Medicaid stimulated the growth and development of these institutions. Before Medicare and Medicaid, a variety of institutions served similar purposes. These were mostly public or nonprofit institutions sponsored by religious groups and included hospitals and homes for the chronically ill or incurably ill, homes and infirmaries for the aged, county homes and infirmaries, convalescent homes. Currently, nursing homes fall into several categories: extended-care facilities designed for short posthospital stays; skilled nursing homes, which focus on long-term intensive nursing and therapy for people with degenerative illnesses, such as Alzheimer's disease, multiple sclerosis, quadriplegia, and so on (similar to chronic-disease hospitals); and intermediate-care facilities, which provide custodial care.

Some critics claim that the growth of these facilities has encouraged families to shift from volunteer care by families at home to warehousing of the elderly in nursing homes. Supply has created demand.

Hospice care is a relatively new service in the United States. It is designed to provide palliative services and pain relief to the terminally ill, mainly cancer patients, for whom active medical treatment is no longer warranted or desired. Services provided include home care, pain relief, social services, and bereavement services for the family and the dying. These services can be provided at home, in the hospital, or in a facility similar to a nursing home. Recently Medicare began providing coverage for these services. While most hospices are still community, hospital, or religious-group-sponsored programs, for-profit interests have entered the market since Medicare began to cover these services.

Free-standing birthing centers have emerged in the last several years. Initially, they were begun by nurse midwives and were subject to major opposition from obstetricians and hospitals. Part of the stimulus for these centers was the denial of privileges to nurse midwives in hospitals and the reaction of consumers to the medicalization of normal childbirth. The refusal of many health-insurance companies to cover services in birthing centers inhibited the spread of these programs. Furthermore, a birthing center requires physician and hospital backup if problems arise. Physicians and hospitals are now developing free-standing birthing centers.

Costs of childbirth are lower, since lengths of stay are shorter, overhead costs are lower, and fees for nurse midwives are lower than for physicians. However, if birthing centers remove uncomplicated cases from hospitals, the cost per hospital case for maternity care will rise.

The Medicalization of Social and Personal Services

Services that historically were viewed as social and personal-care services have increasingly been subsumed as part of the health industry, and pressures to include new services under this rubric continue. For example:

Home health care includes personal-care services such as bathing and dressing. *Intermediate-care* facilities mainly provide personal-care services of bathing, feeding, dressing, and recreation. *Hospice* services include custodial and personal care, social, and bereavement services.

When services are "medicalized," pressures develop almost immediately to accredit employees and set "standards" for employees and facilities; to pay for services formerly provided by families or volunteers; and to upgrade wages to conform more closely to the wages of the health industry. These factors tend to increase costs over time.

Other types of services are at the doorstep waiting to be classified as reimbursable health services. These include: physical fitness, weight reduction, a range of counseling and health-education programs, and homemaking services for the elderly and disabled.

Alternatives to Hospital Care: Cost Saving, Cost Shifting, Cost Additive

There is little question that substitution of less expensive types and sites of care can save money. What is less certain, however, is (1) whether certain types of care actually substitute for higher-cost alternatives or merely increase total utilization, and (2) whether the "unbundling" of hospital care will reduce total expenditures or just shift higher costs to hospitals. Except for a few types of services, most notably HMOs, and sites of care, such as birthing centers, there is insufficient information to reach conclusions.

Costs of surgery in surgicenters are lower than in hospitals. However, there are two unanswered questions. If all minor surgery is moved out of the hospital and the unit costs of in-hospital surgery rise because only the most severe cases are handled there, will total costs for surgery be the same, lower, or higher? If surgicenters increase in number, will the demand for elective surgery rise? The management of elective surgery through prior admission certification and second surgical opinions could serve as a counterbalance.

The same issues apply to urgent-care centers. Another problem is the possible reduction of quality of care because of the fragmentation of primary care and the absence of "case management" in such centers. Almost no information exists today on either the numbers of surgicenters or on the costs of the care they provide.

For diagnostic centers, in addition to questions of shifting a more severely ill patient load onto hospitals and thus increasing hospital unit costs, there are major concerns about quality control, morbidity, and mortality rates, and the effects of the proliferation of expensive high technology on demand and total costs. Here, too, there is a lack of data on the number of centers, costs, and cost tradeoffs.

Utilization control and increased costs and charges to users are growing concerns in relation to home health services. The federal government has proposed copayments under Medicare, and some insurers provide case management for these benefits to control utilization. Extended-care facilities (nursing homes) may have reduced the length of stay in hospitals. However, since increasing the proportion of acutely ill patients in hospitals increases unit costs, the total cost of care may be the same.

For intermediate-care facilities, questions can be raised as to whether these facilities should be considered health-care facilities. The availability of Medicaid payments for support of care in these facilities may have increased demand, shifted responsibility from families, served as a disincentive to develop resources for affordable congregate housing and personal care, and also increased costs because of the high standards of facilities and personnel required in the health-care sector.

There are too few birthing centers yet to have a major impact on hospitals, or on cost. However, some plans, such as Blue Cross of Philadelphia, have started to pay for home health and homemaker services if a new mother returns home within 24 hours after delivery. Plans such as these can stimulate the further development of birthing centers.

Major questions can be raised as to whether hospice services are additions to rather than substitutes for hospital care. If patients have already exhausted all active treatment possibilities, these are additional services. However, there are not yet sufficient data to reach conclusions.

Alternatives to Hospital Care: Issues of Development and Payment

So little is known about the costs and benefits of existing and developing alternatives to hospital care that caution is a key word. Experience has shown that reasonable and careful efforts to control costs in one part of the health industry without controlling other parts—or controlling costs but not utilization—have led to cost shifting rather than lower costs. Except for data on populations enrolled in HMOs and data from a few comparative studies, no studies are available on total costs for controlled populations using alternative services, particularly new services. Studies reviewing the abundance of services and sites have used price of specific service rather than the cost per illness or total costs as the benchmark, and hence do not provide information on total costs or savings.

In sorting out these complexities in an era of very rapid change, the following are issues that should be addressed.

1. Is the service likely to be a substitute or an add-on? Have hospital days, elective surgery, total utilization, and total cost been reduced?

2. Has total technological capacity been increased in the community, raising total costs? Has quality been reduced?

If diagnostic centers and surgicenters are not subject to certificate of need, health-planning reviews, or quality standards, there are several potential undesirable side-effects—for example, excess capacity, which can add costs; underutilization of technology, which requires frequency of use for maintenance of skills; and a consequent increase in morbidity and mortality.

3. Is the proportion of acute-care patients as well as costs rising in the hospital sector? When hospital costs are combined with the costs of some of the alternative services, have total health-care expenditures and costs risen, fallen, or remained the same?

4. Have patient shifting and cost shifting occurred, with the sickest and poorest patients limited in their choice of sites? Has the burden on hospitals, or on the public programs to provide care, increased?

5. Would it be more effective to stimulate systems of care like HMOs, where total costs, utilization, and quality can be monitored, than to promote separate and special centers?

Certain actions that may make eminent good sense in the short term, when viewed from the perspective of an employer, union, or employee, may have profound negative effects for the nation and, ultimately, for employers and employees. In viewing the health scene, the balance between saving money for a benefit plan and long-term societal implications is extremely difficult to achieve.

Cost Sharing for Beneficiaries

A long and acrimonious debate has raged over the different forms of cost sharing and their value. There is a growing trend to increase cost sharing in health-benefits contracts, as they are renewed, and in the Medicare and Medicaid programs. Proponents of cost sharing believe that payment by the insured at the time of service discourages unnecessary utilization and encourages prudent purchasing and careful review of bills by the consumer. Opponents are concerned that utilization of needed and preventive services will be discouraged, particularly for lower-income groups.

Health-insurance benefits with low cost-sharing provisions or none are called first-dollar coverage even when the insured pays part of the premium. Cost-sharing discussions rarely include reference to the fact that in a noncomprehensive benefits package, while there may seem to be low cost sharing or none, actual cost sharing may be high because of noncovered services.

Finally, for plans which do not have a cap on the amount of cost sharing, even a small percentage of coinsurance can lead to catastrophic costs being passed on to the seriously ill and low-wage employees. For example, 10 percent coinsurance on a $50,000 bill amounts to $5,000. But an employer may agree that once an employee has incurred $1,000 of coinsurance in a given year, the remainder of his or her bills are covered in full. In the last ten years, such stop-loss provisions have become increasingly common. While many economists have recognized that cost sharing has differential effects at different income levels, few plans have income-related cost sharing because of the administrative costs of operating such a plan. However, Jones & Laughlin Steel Company and Xerox now relate cost sharing to wages, and some other companies have related premium contributions to wages.

Many innovations are being explored in relation to the use of cost sharing or cash rebates to influence utilization. Medical expense account approaches have been developed, as have rebate programs. Xerox; Quaker Oats; and Aluminum Workers of America, together with Alcoa and Reynolds, have variations on such plans.

Some of the discussion regarding the value of cost sharing has abated since the findings of the Rand study, which found a direct correlation between the existence of cost sharing and utilization, with no perceived effect on health outcomes. However, a recent study of California Medicaid shows the negative health effects of cost sharing for low-income people.

There is little question that cost sharing influences the use of services. The real issues are not cost sharing per se but several other considerations. What effects, desirable and undesirable, will cost sharing have on which groups in the population and on which services? Can cost sharing be tailored positively to encourage (1) the use of certain services, such as preventive services; (2) the appropriate use of services; and (3) the use of appropriate sites for services? Will cost sharing be used to pass costs back to the consumer, prevent needed care, or distort patterns of appropriate care?

There are four basic forms of cost sharing:
- *Premiums*—the share of total premium costs paid by the employee for his (or her) own care and for dependents' care
- *Deductibles*—a flat sum of money, usually between $100 and $500, that must be paid before the benefit plan begins to cover expenditures
- *Coinsurance*—a percentage (10, 15, or 20 percent) of each bill for any service or for selected services paid for by the enrollee
- *Copayment*—a flat nominal sum for each service of a certain kind, for example, two dollars per prescription.

In shifting from low or no cost-sharing plans to plans that increase cost sharing, employee and union opposition can be a major barrier, unless there is cooperation in the planning of the changes and there are trade-offs in benefits or a share of the savings. A number of companies and unions have made innovative changes in benefits and in the use of cost sharing to encourage or discourage types and sites of care. For example, a number of Teamsters union locals impose a penalty in the form of a $100 fee and 20 percent of the first $1,000 if a patient doesn't receive preadmission testing. There is an increasing trend in companies like South Central Bell to impose cost sharing on in-hospital diagnostic and surgical services but not on outpatient services. Rockwell International waives the 10 percent coinsurance for selected procedures performed on an outpatient basis.

Design and Redesign of Health Benefits

When employers and unions move from passive payers for health-care services to active managers of health benefits, one of the first steps is to review and revise the health-benefits program. For example, in response to The Business Roundtable's call to assess business's contribution to slowing the escalation of health-care costs, Pfizer examined its own health-care cost history and developed a two-pronged program. The first part focused on the structure, content, and payment systems of its benefit plans and health-care delivery system. Pfizer concluded, as have other corporations and unions, that while shifting costs from the company to the employee would provide immediate savings, it was not the solution.

A redesign of benefits requires review of current utilization and cost patterns to isolate areas where changes in benefit design can have a positive impact. Among the key factors for review that should be evaluated are:

☐ *Scope of benefits.* Are certain preventive services with known positive outcomes included, such as comprehensive prenatal and postpartum care, immunization for children, and so on? On the other hand, are experimental services with unproven value, which may be very costly in aggregate or per procedure, excluded?

☐ *Options for different sites of care.* Do the current benefits include or preclude the use of less expensive sites for certain procedures? For example, are ambulatory surgery, home health services, birthing centers, and so on, covered services? Do employees have a choice among benefit plans and are HMO/PPO options included in the choice?

☐ *Incentives.* Is cost sharing tailored to discourage unnecessary and inappropriate use and sites of care? Are there incentives to encourage certain types of behavior among consumers, such as preadmission testing, use of preventive services like Pap smears, and so on?

☐ *Side-effects.* Does the amount of cost sharing place such a heavy burden on low-income families or those with catastrophic illness that it discourages necessary utilization or causes economic hardship?

There are a number of other factors that need to be considered in benefits design and redesign. These include options that can lead to adverse selection, plan complexity, coordination of benefits, and payment to providers. Payment to providers will be discussed in chapter 4, "Managing Health-Care Benefits."

Premium levels, cost-sharing provisions, scope of coverage, and incentives may induce low utilizers to enroll in one plan, while the premiums for other plans, which attract those with medical problems, escalate and total benefit costs rise. Thus, companies that offer multiple options need to evaluate carefully the utilization and cost experience of each option separately.

With the increased participation of women in the work force, it is becoming more common for families to have overlapping medical coverages. Unless benefits under the two plans are coordinated, utilization and cost-containment provisions can be negated.

Finally, in the attempt to refine a benefits plan through the use of incentives, variations in premiums, extensive multiple choice, and cafeteria plans, the choices sometimes become difficult for employees and their families to understand. Simplicity might outweigh the advantages of a very complex redesign of benefit plans.

RECOMMENDATIONS

4 Wherever diagnostic, preventive, or treatment services can be provided outside a hospital as safely and effectively as within, and at lower cost, employers and unions should ensure that the health-benefits plan encourages their use by employees, dependents, and retirees. Where such services are not available, employers and unions should stimulate their development.

5 In determining whether a particular out-of-hospital service deserves to be encouraged, employers and unions should take into account such factors as: (1) whether the service unnecessarily increases the total consumption of health care in the community; (2) whether the lower cost of the service has the effect of raising hospital costs in the community; and (3) whether the location of the service causes hardships for the poorest and most seriously ill patients.

6 When employers and unions decide to adopt or increase cost sharing, they should carefully tailor it to encourage cost-effective utilization and to discourage wasteful utilization, but not to cause employees to forgo necessary care. They should also ensure that cost sharing does not place undue burdens on low-wage or catastrophically ill beneficiaries.

7 Employers should involve employees and unions in designing as well as implementing the health-benefits plan. Such participation leads to larger, longer-lasting gains.

REFERENCES

Allen, Richard. "Health Care Cost Management at the Worksite—the Pfizer Approach." An unpublished paper commissioned by Work in America Institute, January 31, 1984.

Fox, Peter D; Goldbeck, Willis B; and Spies, Jacob J. *Health Care Cost Management.* Ann Arbor, Mich.: Health Administration Press, 1984.

Fuchs, Victor R., "The Battle for Control of Health Care." *Health Affairs,* Summer 1982.

Hanft, Ruth S. "Alternatives to Hospital Care." An unpublished paper commissioned by Work in America Institute, May 9, 1984.

Hembree, William E. "Joint Labor and Management Efforts to Control Employee Health Care Costs." An unpublished paper commissioned by Work in America Institute, May 9, 1984.

Light, Donald W. "Is Competition Bad?" *New England Journal of Medicine* 309 (November 24, 1983).

Moxley, John H. III, and Roeder, Penelope. "New Opportunities for Out-of-Hospital Health Services." *New England Journal of Medicine* 310 (January 19, 1984).

Newhouse, Joseph P., et al. "Some Interim Results from a Controlled Trial of Cost Sharing in Health Insurance." *New England Journal of Medicine* 305 (December 17, 1981): 1501-1507.

Relman, Arnold S. "The New Medical Industrial Complex." *New England Journal of Medicine* 303 (October 23, 1980): 963-970.

Rice, Thomas H. "The Impact of Changing Medicare Reimbursement Rates on Physician-Induced Demand." *Medical Care* 21 (August 1983).

Saline, Lindon E. "An Overview of The Business Roundtable Health Initiatives." An unpublished paper commissioned by Work in America Institute, January 31, 1984.

4. MANAGING HEALTH-CARE BENEFITS

Until recently, most large and small purchasers of health-care services did not actively manage their health-benefit programs. The increases in cost and the growing dissatisfaction with the shifting of costs from the public to the private sector have led large employers and unions to intervene in two major interrelated directions: to increase competition and market forces in the industry; and to manage their own health-care benefits through the design of benefits that assure appropriate use of benefits and the payment of reasonable prices. Successful management strategies must be based on knowledge of the complexity of the health industry.

The Business Roundtable recognized this complexity in a pamphlet outlining its approach to health initiatives. It stated:

> The health care system is very complex, technically, economically, socially, politically, legally, philosophically and ethically. The health care system is an anomalous market considering relationships of supply, demand, costs, charges, users and payers.

Health care can never be totally subject to market forces, unlike most consumer products. Providers and users should set prices for health care in a competitive market as far as possible. However, high-quality health care, like education, should be available to everyone, regardless of the ability to pay. Lack of access to care, and delay in preventing illness or in diagnosing and treating illness, can cause lost productivity, disability, and premature death. The social costs in lost economic goods; disability dependency; and increased federal, state, and local taxes are large.

Moreover, the consumer of health care is often not the direct purchaser of care. Because of the knowledge required to make decisions on appropriate patterns of diagnosis and treatment and on quality, the provider of care, usually the physician, acts as agent for the consumer. Accreditation, credentials, and licensure limit the entry of providers into the market. In addition, the extensive third-party payment system of "health insurance" removes most economic considerations from the consumer of services at the time of service.

Until very recently a perceived shortage of physicians limited the ability of consumers, third parties, management, and labor to manage health benefits actively because they feared loss of access to care by consumers. However, the doubling of the number of physicians trained in the last 20 years, the overcapacity of the hospital sector, and the entry of numerous alternatives to hospital care have led to an environment of competition among providers. Active management of benefits by the private and public sectors can now make a major impact on the utilization, quality, and cost of care. Physicians and hospitals rarely negotiated price in the past because the purchaser paid the charge without question. Today, health maintenance organizations (HMOs) and preferred provider organizations (PPOs), which control or negotiate costs and prices, are developing rapidly.

The preceding chapter described how health benefits can influence use, quality, and costs. While the appropriate design of health benefits can help influence utilization, quality, and cost of care, benefit plans must be closely monitored, managed, and evaluated to assure that the objectives of the plan are met and to modify provisions of plans that are less than effective.

HOW TO SELECT POLICIES FOR MANAGING A HEALTH-CARE PLAN

Health benefits can be managed with selected policies and techniques, targeted at identified problems, or with a comprehensive program of activities incorporating the following:

☐ *Claims and utilization review*, to identify problems and to

monitor accuracy of billed services and charges as a basis for designing utilization, quality, and cost controls
- ☐ *Utilization controls*, such as prior authorization programs, concurrent review, second surgical opinions, and standards and criteria respecting use of services and sites of service
- ☐ *Quality assurance*, to limit choice of providers to those who meet preset standards; includes peer review, morbidity-mortality and medical-evaluation studies, and small-area variation studies
- ☐ *Prudent-buyer strategies*, such as selection of providers and sites of service on the basis of quality, accessibility, and price; stimulation of the development of alternative delivery systems and competition; and reimbursement policies
- ☐ *Education* of employees, their families, and managers concerning the influence of individual behavior on health use and cost, price comparisons, quality considerations, and alternative sources and sites of care.

Since data are key to the management techniques described, available data sources are reviewed.

Management techniques for influencing, monitoring, and managing health care can be cost-effective. However, there are administrative costs related to techniques of management, data collection and analysis, and implementation of utilization controls.

Before employers and unions begin to collect and analyze large amounts of new data and to implement programs, they need to define why they want to proceed; what they will use information for, once it is analyzed; and the costs of utilization and other programs. For example, if a major corporation and its union have identified a specific problem of high costs in one location versus others and want to bring this location into line with others, it would be useful, before instituting a utilization control program, negotiating with providers, and so on, to have the following types of information:

- ☐ Are there major differences in the age, sex, or race distribution of employees and dependents?
- ☐ Are there major differences in the medical market-basket costs in the area?
- ☐ Is there a high incidence of an occupation-related occurrence?

The next steps would be to review claims data (assuming the data have some detail) as follows:

56   *Improving Health-Care Management in the Workplace*

- ☐ Are claims routinely reviewed for accuracy? If not, what are the error rates on bills, and will rigorous claims review resolve most of the problems? Do the claims forms show enough detail to be able to determine the use of ancillary services, for example?
- ☐ Are there major differences in admission rates and lengths of hospital stay among the different plant locations? If so, preadmission certification and concurrent review, or changes in cost sharing, might reduce the differences.
- ☐ Are there major differences in surgical rates and sites for elective surgery, such as biopsies, hysterectomies, tonsillectomies? If so, in high-use areas, a second surgical opinion program may be warranted for specific types of elective surgery.
- ☐ Are too many employees using high-cost providers when low-cost providers are available?

None of the above requires very sophisticated data collection and analysis. It does require adequate claims forms and diagnostic and procedure data. On the other hand, if an employer wants to sponsor or participate in a PPO and to select and negotiate with hospitals, physicians, or HMOs in the area, a lot more provider-specific information may be needed; the assistance of a carrier may be advantageous in all these respects. Quality of care becomes important. For example, do the morbidity rates for specified types of surgery differ in different hospitals or surgicenters in the region? However, this is not easy data to acquire.

The most important first step is to determine the key purposes to be addressed by a management program. For example:

- ☐ Is the issue that total costs for the employer and union have risen faster than average national, state, or regional costs and that a comprehensive company-wide plan is desired?
- ☐ Is the issue that the company or union wants to slow the growth of its own health-benefit costs even if costs continue to rise more rapidly regionally or nationally?
- ☐ Is the issue that plant A or local A is incurring utilization rates and/or costs much higher than other sites?
- ☐ Has a particular problem already been identified, such as high hospital admissions or surgical rates, that requires further analysis and, more important, design of control mechanisms?

The approach to designing a management program, data collection, or analysis will be different for each of the above.

Time should also be spent at the beginning in considering whether there are any future actions that, even if they could affect the problem, would not be taken for practical reasons, such as union-management relations or recruitment. For example, negotiations over price with providers may not be a viable alternative in the absence of a strong business coalition or a consortium of large purchasers.

On a related point, it is obvious that changes will affect employees and management—for example, changes in the way they receive benefits, in the value of the benefits, and in access to care. If incentives and penalties are contemplated to ameliorate these problems, then the best first step for success is early involvement of managers and employees in reviewing the changes, reviewing the supporting data, offering ideas for solutions, and designing the management response.

Good claims data can be used for multiple purposes—for example, analysis and control of utilization patterns and costs, cross-plant comparisons, and the like. If you already have a preadmission certification program, is it working? If not, why not? Can one be implemented quickly, or is it necessary to wait for contract negotiations? Are there organizations in the area which have experience operating utilization control programs?

There is a considerable body of experience with claims review preadmission certification, concurrent review, second surgical opinion, and peer review programs, and with the incentives and disincentives paired with these programs. Each program should be analyzed with respect to review criteria, administrative options, and costs.

Last, it may be feasible to obtain needed data through relatively simple expedients, such as:

☐ Changing claims forms to provide the information needed
☐ Convincing the carrier or the administering insurer to provide data (other than data about individuals) from other accounts
☐ Convincing employers and unions in the area to pool data
☐ Convincing the state to sponsor a statewide data-collection system
☐ Seeking advice from firms experienced in collecting and analyzing data
☐ Acquiring review systems such as the "Appropriateness Evaluation Protocol" or other standardized criteria or review programs.

A COMPREHENSIVE MODEL FOR MANAGING HEALTH-CARE BENEFITS

One of the most comprehensive approaches to managing health-care benefits was developed in 1984 by the Midwest Business Group on Health. It proposes a fundamental change in the way private health benefits are managed. The model system contains seven interdependent elements:

- *Selection of effective and efficient providers*, based on factors that include quality criteria from quality-assurance programs, mortality rates adjusted for severity of illness, proficiency levels of physicians in selected specialties, and up-to-date diagnosis and treatment protocols.
- *Hospital payment* based on prospective rates negotiated in advance, and on the encouragement of use of the most appropriate settings.
- *Physician payment* based on fee schedules which encourage conservative practice and choice of inexpensive settings.
- *Payment of other providers* on the basis of prospective payments for alternative sites of care and delivery systems; encouraging the organization of physician groups into freestanding alternative centers that could be affiliated with selected hospitals.
- *Utilization management* to ensure that treatment is necessary, cost-effective, and of high quality, and that it provides clinical information with which to track physicians' practice patterns and to educate providers.
- *Monitoring and evaluation* based on accurate and timely data about diagnosis, quality of care, and demographics.
- *Plan design and communication*, including education programs for employees, and incentives and disincentives to promote cost-effective purchasing.

The chart on the facing page summarizes this plan.

While the plan presents a comprehensive approach to purchasing health benefits, implementation for many employers and unions will take time and will be dependent on obtaining information on use, quality, cost, and so on. While all employer/union plans can adopt parts of the management strategy, the ability to negotiate with providers will depend on the size of the health-benefits group, the supply of providers, and competition in the area.

Competitive Health-Care Purchasing System Model System Compared to Current Health Insurance Benefit Payment System

Element	Current Primary Health-Plan Payment System	PPO/HMO System	Competitive Health-Care Purchasing System
Selectivity of providers	All licensed providers are included	Self-selected providers	Providers chosen, using quality, cost, and location criteria
Hospital payment	Line-item charges at current prices	Line-item charges at discount prices	Prospectively determined prices, per diem/per case, or other plus bonus
Physician payment	Usual, customary, and reasonable charges	Prospectively determined charges, capitation, or salary plus bonus	Prospectively determined fee schedule plus bonus
Other providers payment	Line-item charges at current prices	Charges at discount prices	Prospectively determined per-case price plus bonus
Utilization management	Some pre-admission and concurrent review	Comprehensive utilization review	Comprehensive review and case management
Data base and monitoring	Claim-based records, standard reports	Some encounter records and summary reports	Integrated episode record and flexible management information system
Plan-design linkage	Same benefits for all providers	Full coverage for selected providers	Reduced coverage for nonselected providers
Employee communications linkage	Communicate plan benefits and limits	Communicate plan benefits, service locations	Communicate prices, services, and locations of selected providers

INDIVIDUAL PROGRAMS AND SUPPORTING DATA

Supply Data

Supply data are important in choosing interventions that involve providers of care. If there is only one hospital in an area, or the area is a designated health manpower shortage area, the ability to intervene on the provider side is very limited. Sources of data on providers include the following:

☐ The area resource file (maintained by the Department of Health and Human Services) merges data on health-care facilities and manpower by standard metropolitan statistical area. It is a prime source for consulting firms that conduct strategic planning or marketing activities in health. County data on physicians are also available.

☐ Data on hospital use are available from the American Hospital Association surveys. These surveys provide considerable detail on the type of hospital, scope of services, staffing, and so on.

☐ The American Medical Association's socioeconomic monitoring survey collects and publishes data on physician activities, charges, and income, by region.

Claims and Utilization Review

Rigorous review of claims and utilization is the most common basis for identifying problems with benefits design and the management of benefits, use, and costs of care. Much of the data needed for initial productive interventions can be obtained from good health-insurance-claims data or direct-payment data, supplemented by national and regional data banks. States such as Iowa and Illinois are beginning to develop statewide data systems on health-care use and costs, using a uniform hospital bill designed specifically to assist employers, unions, consumers, and third parties in designing and operating cost-effective benefit programs.

Claims data, depending on the size of the data base (employer-specific, or from multiple employers) and the design of the claims form, can be used for a number of purposes:

☐ Review of the claims for billing errors. There is considerable evidence of high billing-error rates. All bills should be reviewed by employers or the insurance company and employees for accuracy.

☐ Analysis of utilization patterns within the firm, and within the immediate geographic area among different employers, controlling for age and sex differences. These data can quickly identify conditions truly outside the norm, such as high hospital admission rates, above-average lengths of stay and surgical rates, and high usage for certain types of services. More detailed data, arranged according to major diagnostic categories, can be profitably analyzed—with the caveat that unless diagnostic related group (DRG) analysis of hospital data is undertaken, there will be data comparability problems.

Provider patterns can also be identified if the claims data contain good identification of providers. Statistical profiles on provider-specific utilization can be developed. However, patients have to be classified into clinical groups, for example, diagnosis, age, sex. DRG analysis is a good tool for comparison, although it doesn't account for severity of illness. Pooled claims data covering a number of employers in an area can be a powerful tool for managing health benefits and for educating consumers and providers. The availability of pooled data will vary from area to area.

Data that are specific to a small employer or small union local will be only marginally useful for benefit design, utilization and cost control, or negotiation with providers. Furthermore, in areas with numerous competing insurance plans, the pooling of data may be impeded by differences in claims forms, diagnostic coding, and the proprietary interests at stake.

A number of employers and insurance companies have begun to design and install claims processing systems which specify not only the data that must be furnished by providers, but the standardized format to be followed. For example, Deere and Company has a system called Comprehensive Insurance Claims Handling (CINCH), which requires the following information:

—diagnosis code by ICD-9-CM (*International Classification of Diseases, Modified*, 9th ed.)
—procedure code by CPT-4 (*Current Procedural Terminology*, 4th iteration)
—provider-of-service code
—patient data (name, address, Social Security number, age, sex)
—claim-type code (hospital, medical, dental)
—dates of treatment
—amounts charged
—amounts denied and reason code
—amounts paid and date paid
—Medicare eligibility
—coordination-of-benefit information.

As claims and utilization data become standardized and increasingly available for comparison, small-area variation studies can be undertaken and used both to select utilization-control mechanisms and to influence provider behavior. Pioneered by Dr. John Wennberg of Dartmouth Medical School, studies of variations in medical practice among communities in northern New England have been used to identify aberrant patterns of surgery, laboratory tests, and so on. These data also have been used to influence changes in physicians' practice patterns.

For data other than individual benefit-plan data, the following are sources:

☐ Carriers and intermediaries are still the primary source of claims and utilization data for employers and unions. Recently they have been establishing relationships with data firms that can develop industrial and area profiles. The Health Insurance Association of America, in response to coalition requests, recommended a standard set for hospital inpatient data; unfortunately, the recommendations on diagnoses are not compatible with DRGs. Blue Cross, Corporate Health Strategies, Metropolitan Life, and the Health Data Institute have worked to improve profiles. There are also third-party administrators who are active in claims and utilization review. One of these, U.S. Administrators, has developed computerized models.

☐ Business coalitions have stimulated the creation of data foundations and state data centers in their regions to provide pooled information on utilization and costs. Examples are Philadelphia, where comparative hospital data are generated, the Utah Health Cost Foundation, and the Health Policy Corporation in Des Moines, Iowa. The latter, a quasi-public body, collects data on all hospitalization in the state based on a uniform bill (UB82). The Fairfield-Westchester coalition in New York and Connecticut has established a data consortium. The Midwest Business Group on Health works with insurance carriers to develop uniform data elements. The Minnesota Coalition on Health Care Costs helped stimulate community-wide review; currently 16 major companies in the Minneapolis/St. Paul area have contracted for concurrent and preadmission review. In addition, numerous private organizations provide extensive data analysis service.

☐ Professional review organizations (PROs) and their predecessors, professional standards review organizations (PSROs), frequently expanded beyond Medicare and provided services to large insurers and companies, although their main purpose was to review utilization of Medicare patients. Many of them have useful data on utilization patterns. A number of these organizations also conducted special evaluation studies. Du Pont and Caterpillar use PRO programs.

☐ A number of state mandatory rate-regulation commissions, such as those in Maryland, Massachusetts, New York, and New Jersey, collect and analyze data on admissions, occupancy, and utilization patterns within their states, as well as cost data for specific institutions and, sometimes, for specific diagnoses. Some states with voluntary commissions, such as Virginia, also conduct annual cost and utilization surveys of the most common diagnoses.

☐ Data tapes are available from several National Center for Health Statistics (NCHS) surveys, with information on national and sometimes state data. These include Hospital Discharge Surveys, Health Interview Surveys, National Ambulatory Medical-Care Surveys, Medical-Care Use and Expenditures Surveys. Data from these surveys can be used to compare local experience with national and sometimes regional patterns, by age and sex. Incidence and prevalence of illness data are also available. Annually, NCHS publishes a document entitled *Health, United States*, which compiles data from many sources on health status, health-care resources, and health expenditures.

☐ The Health Care Financing Administration (HCFA) provides utilization data on the over-65 population. These data can be useful in analyzing utilization patterns for retirees. The Medicare cost reports also provide cost data discussed later. In addition, every hospital has a Medicare severity indicator which helps to distinguish hospitals by the complexity of services provided. HCFA also compiles the National Health Expenditures data, including per-capita costs by type of service.

☐ The American Hospital Association conducts annual surveys of hospitals and publishes data on their characteristics, occupancy, volume, revenues, and expenses, in addition to regional utilization data.

Utilization Control Mechanisms

Utilization review and control are not new techniques. The original Medicare legislation called for utilization committees in each hospital to review admissions and lengths of stay. Utilization review was expanded with the establishment of professional standards review organizations and now by professional review organizations. Most of the larger HMOs, such as Kaiser, have internal review processes, along with protocols or criteria for hospital admissions, lengths of stay, and elective surgery.

Medical care continues to be part science and part art, with wide variations in diagnostic and treatment patterns for similar conditions. Due to the absence of rigorous technology assessment and rapid advances in scientific knowledge, there are large areas of uncertainty in medicine and justifiable differences in practice patterns. However, certain utilization control tools can affect average admissions, lengths of stay, elective surgical rates, and use of laboratory and X-ray services.

☐ *Profile analysis*, an extension of claims and utilization review, focuses on provider-specific patterns by age, sex, and patient diagnosis, and DRG analysis of inpatient hospitalization. DRG review analysis is being sold to private companies and insurers. Other profile review systems are also available—for example, the "Appropriateness Evaluation Protocol" (which requires assigning personnel to hospitals to determine the necessity of admission). These data can be used to identify providers whose patterns of care appear to deviate from norms, and to educate them, deny them participation in certain programs, or make participation subject to prior approval.

The data bases for these reviews often present problems, for example, medical records contain inaccuracies, such as differences between admission and discharge diagnoses, errors in coding, and so on.

☐ *Preadmission certification.* A rapidly growing number of health-benefit plans require certification before an enrollee can be admitted to a hospital for nonemergency services. Preadmission certification can apply to permission for the actual admission, for the projected length of stay, and for the use of high-cost diagnostic and treatment procedures. Companies that require preadmission certification include Hospital Corporation of America, Westinghouse, Mitre, Levi Strauss, Bank of America, and Deere and Company.

Responsibility for preadmission certification can be delegated to employees of a hospital or to private nonhospital organizations. The Hospital Corporation of America believes there is a so-called "sentinel effect," that is, physicians and patients alter utilization patterns in order to avoid the disapproval of admissions. Most health maintenance organizations are successful in large measure because they reduce hospital utilization and specialty services through formal programs of preadmission certification and prior approval. In its broadest context, preadmission certification subsumes second-opinion surgical programs. It can also include prior authorization for nursing-home care, home health services, specified high-technology services, and specified providers.

The program most frequently discussed is that of the Minneapolis Foundation for Medical Care Evaluation, which reviews admissions for more than 100 companies in that state. The criteria for review are adapted from the "Appropriateness Evaluation Protocol."

Preadmission certification can add substantial administrative costs, depending on the scope and type of review, but these are often more than offset by the savings.

☐ *Concurrent review* examines the appropriateness and medical necessity of hospital admissions and stays, while the episode of illness is in progress. Although professional standards review organizations pioneered this type of review, it is usually conducted by nurses, using medical records based on diagnostic-specific criteria. At present, significant cost savings appear to accrue from concurrent review programs. However, as they become more widespread, and as rates of hospital admissions and lengths of stay decline, the cost-benefit ratios will change.

☐ *Preadmission testing* has requirements that are more controversial than other forms of utilization control. Because of concerns about quality, many hospitals and specialists will not accept tests performed in another facility or by another practitioner. Also, it is difficult to monitor whether a test is or is not repeated. However, the pressures to control utilization have led to changes in hospital policies, so that consumers now can have the needed tests taken at the hospital just prior to admission, without incurring costs for an overnight or weekend stay.

☐ *Second surgical opinions* were originally initiated in response to the concerns of health leaders that surgical rates were too high and that unnecessary surgery was taking place. There was the additional fact that almost every surgical procedure carries risk to the patient. Initially, these programs focused on whether there was need for the type of surgery ordered. Today, they also include authorization for the site of surgery (inpatient or outpatient). Most have gradually narrowed the types of surgery requiring review to procedures which have high-use patterns or are subject to wide variations in use rates among similar populations.

Most second-surgical-opinion programs are voluntary; a small but growing number are mandatory. Programs pay for the second or even third opinion and for associated laboratory and X-ray procedures. However, if a program is voluntary, with no incentives to follow the advice, it can be cost-additive. Some employers have begun to apply cost-sharing penalties if a second opinion is not obtained for selected procedures.

Cost-effectiveness of these programs depends on their design. Fox and Goldbeck assert that voluntary programs "are likely to achieve only minimal cost savings due to low participation rates and fixed costs for program operation."

Quality of Care

Measurement and assurance of quality of care have always been problematic, for several reasons: only minimal agreement on standards of practice; limited methodologies for evaluation; wide variability in practice and utilization patterns, reflecting the uncertainty of much medical care; scarcity of assessment studies on the efficacy of treatment technologies; and problems of confidentiality with individuals' medical care. Many quality measures are gross measures of process rather than of outcome.

☐ *Accreditation of institutions.* Hospitals and certain outpatient facilities are accredited by the Joint Commission for the Accreditation of Hospitals (JCAH); the standards are regarded as minimal. However, denial of accreditation or probationary status signals real problems with an institution or part of an institution. JCAH also requires that hospitals perform medical evaluation studies.

☐ *Certification of personnel.* A physician, once licensed, is relatively free to provide any physician's service. However, each specialty certifies physicians who are trained in that specialty and who pass a board examination (board certified). Again, this is a one-time examination, except in the specialty of family practice, which requires the physician to be recertified.

Other, more subjective peer criteria are used by physicians and others in selecting physicians. For example:

—Whether the physician is a graduate of a U.S. or Canadian

school or of a foreign medical school, and the country in which the foreign school is located. For example, training in England, Israel, and the Scandinavian countries is well regarded; training in all of the Caribbean, most of eastern Europe, and the Middle East, in some Mexican schools, and in certain institutions in Italy is problematic.
—Where the physician completed his or her residency.
—The physician's teaching status and hospital affiliations.

Clearly, there are signals of quality problems if a physician has lost licensure in a state and moved to another (the information is difficult to obtain, of course) or has lost hospital admitting privileges for disciplinary reasons.

On the other hand, the certification of allied health professions and attempts to restrict to single professional groups the right to provide particular services may bear little relation to quality. The nurse midwife/obstetrician and ophthalmologist/optometrist battles exemplify the harmful use of credentials.

☐ *Volume of services provided.* For most types of surgery, and for certain diagnostic and treatment procedures, the volume of services provided by the institution and the individual practitioner can be taken as a proxy for quality. The classic example is open-heart surgery. Types of services in which a minimum frequency of performance is needed in order to assure quality include: most surgery, cardiac catheterization and other invasive diagnostic procedures, neonatal intensive-care services, shock trauma, high-risk maternity cases, certain pediatric services, and so on.

☐ *Morbidity and mortality data.* Mortality data are, as a rule, available. Morbidity data are rarely available, except for national and state data collected by the National Center for Health Statistics and the state health departments. Occasionally, health-planning agencies collect such data, or special medical-care evaluation studies are sponsored by hospitals, universities, or PROs. For example, several years ago morbidity and mortality rates were developed for coronary bypass surgery in the Washington, D.C., and suburban Washington hospitals, with surprising findings; the two hospitals that clearly had the best results were not university-owned, although both were teaching hospitals. PROs have collected these data often, but the results are not widely available because of confidentiality requirements.

☐ *Epidemiological data and studies.* Just as there are wide variations in use and cost across the country, there are similar variations in disease incidence and mortality by age, ethnicity, race, sex, and geographic region. These differences, although the sources of the differences are not clearly known, must be taken into account. Some differences relate to the epidemiology of the population. For example, spina bifida is most prevalent in the Appalachian region among groups of English, Scotch, and Irish descent; hypertension rates are higher for blacks than for whites; New Jersey's cancer rates are the highest in the country. Other differences, as research by Drs. John Wennberg and Philip Caper has shown, relate to practice styles of physicians.

In addition, such occurrences as influenza epidemics can make major differences in year-to-year utilization patterns of physician visits and hospital admissions, and (it is strongly suspected) have secondary effects on infant mortality and on morbidity/mortality rates for pneumonia and cardiovascular illness.

Small-area analyses, initially conducted as health-services research projects, have expanded. They are epidemiologic studies to determine whether differences observed in use of services among populations are due to random variation or patterns of practice. Hospital-claims data form the basis for these studies. The studies by Wennberg and his associates show appreciable differences in per capita rates of hospital admission for surgical and other procedures among hospital service areas. They also indicate that the relative variation reflects the degree of consensus among physicians concerning the criteria for diagnosis and selection of treatment—a measure of professional discretion.

The importance of these quality data is that they can be used to generate valuable information to enable business, employees, and physicians to better manage health-care resources. These data can assist in designing and implementing programs of preadmission certification and second surgical opinions, and in selecting providers.

Cost and Price

Meaningful and comparative cost/price data are essential to containing costs, selecting providers, and negotiating with providers. These data are more difficult to acquire than utilization data, for a number of reasons, including:

—Differences in accounting systems of institutions

—Lack of uniform billing and discharge data for hospitals and

the use of different forms by different insurance companies
—Failure to use uniform diagnostic and procedure codes, particularly for out-of-hospital services. Most hospitals code, but not all carriers require use of the ICD-9-CM; nor is the CPT-4 procedure code used universally.
—Difficulty in gaining access to physician profile/screen information
—Errors in hospital and physician records and in coding.

However, even with these difficulties, certain data sources can be helpful. If there are pooled claims data, they can be used to compare age- and sex-adjusted per-capita costs for employers in the same area. They can also be used to compare volume of specific services or procedures. For employers with multiple sites, per-capita costs can also be used to identify problems in specific sites.

Where the claims data include good diagnostic and procedure coding, and where provider profiles are available, data can be analyzed and used as the basis for education or negotiation with providers and consumers.

The Medicare cost reports provide gross data for comparison among hospitals. DRG specific information is obviously more useful for detailed analysis of costs among providers. However, Medicare cost reports can provide data on revenues or costs per case. These types of data can also be used to compare differences between costs and charges. Medicare also publishes per-capita cost data by state for Medicare recipients, which can be useful in analyzing cost patterns among retired employees.

State rate-setting commissions, whether mandatory or voluntary, have data on individual hospitals and can provide comparative data. They usually require uniform billing forms. Frequently, they perform special analyses of costs and charges by most common diagnoses. In New Jersey, comparisons based on DRGs are available. Blue Cross in Kansas and other states, as well as some insurance carriers, are now using DRGs and should also be able to provide such comparisons.

Information on the charging patterns of individual physicians is the most difficult to acquire, particularly on a physician-specific basis, unless claims forms are set up to identify individual physicians and there is good standardized coding of procedures. Most large third-party payers maintain physician profiles and have data on the range of usual and customary charges, as well as prevailing charges, in a medical market area. However, there are problems with commonality of definitions and coding, which make some of these data difficult to use.

In negotiating with providers for a preferred provider arrangement, it is imprudent to make price the only criterion. Quality and access are necessary to assure acceptability to the consumer and to prevent underutilization. The Medicaid mills are a classic example of the negative consequences of exclusively price-driven approaches.

Problems with Data

Some caveats should be kept in mind when using available data to design benefits, establish utilization control mechanisms, negotiate with provider groups, and develop cost-reduction programs.

Very few small-area studies have been based on samples large enough to analyze more than a few variables. Often reliance has to be placed on comparisons with national or state rates of use and cost, or with company- or union-wide data. The trend to merge claims data bases, fostered by business coalitions, and the development of data banks, such as those in Utah and Iowa, should enhance the ability of companies, unions, and researchers to conduct detailed analyses and small-area studies. Many employment units, however, will still be too small to provide a statistically valid basis for comparison with other employment units, or for local, state, or national comparisons.

There are numerous problems in comparing data from one source to another because of variations in definitions, accuracy of data, sample sizes, and descriptor codes for diagnosis and treatment. The federal government has imposed standards for comparability of data and definitions and has mandated uniform diagnostic codes for data collection for Medicare and other federal beneficiaries, and for vital statistics data and surveys. Nothing comparable exists, however, in the private sector or at the state level.

Quality of data varies widely. In a number of studies of the reliability of medical records and hospital discharge abstracts, high error rates were found.

Certain demographic and epidemiological factors, particularly age, sex, race, and income, have substantial influence on use rates. More subtle characteristics, touched on in the section on epidemiology, also can affect use and costs of care. Analysis must take these factors into account.

Medical information is more sensitive than any other kind. Over time, substantial barriers have been erected to protect against access to individual medical information. Misuse of such information—for example, in hiring or termination—could totally undermine the future ability to collect and analyze such information.

Certain information on providers, affecting the financial viability and market share of the provider, is also regarded as proprietary. Providers will continue to resist collection and/or disclosure of such information.

Educating Managers, Employees, Dependents, and Retirees

Most employees, dependents, and retirees have had inadequate knowledge in the past on the use of services, costs and quality of care, and provider-specific data. Public disclosure and good information can influence their choice for the better. The kinds of data that could be useful for this purpose include:

- □ *Use data.* The patterns of use of health care in their own workplaces and in other workplaces in the company; similar information for their geographic area.
- □ *Cost data.* Real cost data have never been available to employees, dependents, and retirees. These data fall into two categories:
  - —Cost of individual services for different providers; total costs of care among different delivery systems that are offered employees, retirees, and dependents.
  - —Their own annual costs of care; the costs for their dependents; the average costs for the group, company-wide, and, if possible, area-wide.
- □ *Quality information.* Employees should receive the following kinds of information on quality of care:
  - —Generally available community information from health-planning agencies; special studies, such as mortality rates for surgery among institutions.
  - —With respect to HMOs and PPOs, detailed information on the types of hospitals used, and available information on those hospitals; information on the background and training of physicians.
  - —Information on the findings of programs of utilization review, medical evaluation, second surgical opinions, preadmission certification, and concurrent review.

The above discussion focused on the initial analysis of a company's or union's experience and the design of programs to manage benefits. If there is a broader objective to negotiate with providers, such as physician groups, hospitals, and HMOs, then far more complex kinds of data are required, for example, information on:
— Accreditation, credentials, and admitting privileges, for hospitals, physicians, and other providers
— Scope of services provided and volume, for certain types of services; availability of specialized services
— Quality, including medical evaluation studies, mortality and morbidity analysis
— Costs and charges, preferably by diagnosis for hospitals, or cost per case; severity of illness; physician and other provider charges
— Internal quality-control mechanisms for HMOs, PPOs, hospitals, and other providers.

RECOMMENDATIONS

8 In order to derive full value from the health plan, employers and unions should:
— select providers on the basis of efficiency, effectiveness, and quality
— pay providers on the basis of prospective rates, negotiated in advance
— reduce the delivery of unnecessary services or services that could be provided more cost-effectively
— monitor and evaluate plan utilization with timely, accurate data
— educate beneficiaries to use the plan cost-effectively

9 When selecting policies and techniques for management of the health plan, employers and unions should:
— state clearly the goals they wish to achieve, the problems they wish to solve, and the financial and political constraints that bind them
— review the tools already at hand within the company and the community

—examine the costs and benefits that others have experienced with policies and techniques of the types under consideration
—determine what kinds of data will be needed and whether they are available

10 Obtaining cost and performance data for use in managing the plan may be difficult and expensive. Before seeking such information, employers and unions should determine that it is truly essential to their purposes. They should use their clout in the following ways:
—change claims forms to provide desired information
—get the carrier or administering insurer to furnish data from other accounts
—get other employers and unions in the area to pool data
—get the state to sponsor a statewide data collection system
—seek advice from firms experienced in collecting and analyzing data
—acquire standardized criteria or programs for review

11 Employers and unions should ensure that employees, dependents, and retirees receive information on the kinds of health care available and in a form that will help them become prudent consumers. Such information pertains to:
—availability of health services in the community (in their own workplace and in other companies in the community)
—cost (costs of particular services by particular providers; annual costs of care for each employee, for the group, for the workplace, for the community)
—quality (of services delivered by particular physicians, hospitals, and alternative centers in the community)

REFERENCES

American Hospital Association. *Hospital Statistics.* Data from the American Hospital Association Annual Survey. Chicago, Ill.: American Hospital Association, 1983.
American Medical Association, Center for Health Policy Research. *Socioeconomic Monitoring System.* Chicago, Ill.: American Medical Association, Center for Health Policy Research, 1983.
Business Roundtable, The. *An Appropriate Role for Corporations*

in Health Care Management. New York: The Business Roundtable, 1982.

Caper, Philip. "Health Care Cost Containment and Patterns of Medical Practice—The Role of Business." An unpublished paper commissioned by Work in America Institute, May 9, 1984.

Fox, Peter, Lewin and Associates. Discussion with author.

Fox, Peter D.; Goldbeck, Willis B.; and Spies, Jacob J. *Health Care Cost Management: Private Sector Initiatives.* Ann Arbor, Mich.: Utilization Review Health Administration Press, The University of Michigan, 1984.

Hanft, Ruth S. "Data and Management Information for Monitoring Effective Use of Health Care Services and Cost Containment." An unpublished paper commissioned by Work in America Institute, January 9, 1985.

King, Steve, Midwest Business Group on Health. Discussion with author.

Midwest Business Group on Health. "Model Competitive Health Care Purchasing System." An unpublished paper commissioned by Work in America Institute, January 9, 1985.

Saline, Lindon E. "An Overview of the Business Roundtable Health Initiatives." An unpublished paper commissioned by Work in America Institute, January 31, 1984.

Serverino, Frank, Health Policy Corporation, Des Moines, Iowa. Discussion with author.

Singer, Peter, Utah Health Cost Foundation. Discussion with author.

West, O. David. "Pre-admission Certification—A Golden Opportunity to Start Managing Health Care." An unpublished paper commissioned by Work in America Institute, May 9, 1984.

U.S. Department of Health and Human Services, National Center for Health Statistics. *Health, United States, 1983.* Washington, D.C.: National Center for Health Statistics, U.S. Government Printing Office, 1984.

5. WORK-SITE WELLNESS PRO-
GRAMS: FAD OR FUTURE?

Health experts have long criticized the heavy emphasis in health care placed on treatment for acute illness and high-technology medicine, in contrast to activities that stress prevention and early intervention. The public also has become increasingly aware of the personal, behavioral, and environmental factors that influence health. Consumers have increased their interest in sports and physical-fitness programs, changed their diets and drinking patterns, become more active in environmental protection activities. Health professionals have also developed greater interest in prevention and wellness. The newer forms of health-delivery systems, health maintenance organizations, market on the basis of their emphasis on prevention, early care, and avoidance of unnecessary hospitalization and use of medical procedures.

From the perspective of the employer and the union, prevention can reduce long-term health-care costs, reduce disability, reduce absenteeism, and increase productivity and morale. As employers and employees have recognized the potential economic and social effects of disease prevention and health promotion, work-site wellness programs have rapidly developed.

In California alone, a recent survey showed that 78 percent of 424 companies had wellness projects of one kind or another. The increase in these programs is in response to a number of factors, including federal stimulation of health-promotion programs, employer concerns over rising health-insurance premiums, the general public consciousness of potential effects of life-style on health status, the enhancement of what has been labeled "corporate culture" and esprit, or combinations of these.

These "wellness" programs may range from one or two projects targeted at specific problems to a melange of projects, such as health education, smoking cessation, hypertension screening and control, multiphasic screening, nutrition, occupational health and safety, and employee-assistance programs (for substance abuse and mental-health problems). Employee-assistance programs, in many instances, were started before the other programs and are widespread. Wellness programs may be available on site or subsidized in off-site locations. They may be provided directly by the employer's staff, by a contractor, or through a variety of voluntary community agencies. On-site programs may take place in already existing space or may result in capital outlays for building and equipment or in renovation or redecoration of work-site areas.

For a number of reasons, many of the programs have no clearly stated objectives or expected outcomes. The combination of the wide range of potential projects and the lack of clear-cut objectives or outcomes make evaluation of these programs difficult. In fact, by and large, little systematic scientific evaluation of these programs has occurred, and much of the support and promotion of some of these programs appears to be based on less than satisfactory scientific evidence of the interaction between behavior, health status, and costs. For example, considerable scientific controversy exists in relation to the effects of certain types of physical activity, nutrition, and stress reduction on health status, yet these programs are growing rapidly. However, active efforts are now being made to set quantifiable objectives for the programs and to measure outcomes.

Evaluation of a number of the "wellness" programs is handicapped by the time lag between an intervention and a measurable change in health status, or the converse: behavior that may contribute to illness in later life and which is not identified until the employee has left the work force. Even more difficult is the problem of separating the effects of multiple interventions.

There are also methodological difficulties in quantifying costs and benefits, which include such issues as defining costs and benefits uniformly and specifying to whom the costs and benefits accrue; using appropriate and uniform discount rates; measuring quality of life rather than just productivity. Furthermore, certain programs, such as early detection, may result in increased short-term costs.

While the foregoing discussion may appear skeptical about the value of "wellness" programs, there is clear evidence that in some instances these programs are beneficial to employees and to employers. These are generally the programs that fall into the category of disease control, such as hypertension or glaucoma control, immunization, and certain occupational health and safety programs. (The latter are not included in this discussion.)

Less convincing evidence is available on the success of specific smoking-cessation and alcohol-abuse programs, even though there are clear relationships between smoking, cancer, and cardiovascular diseases; between excessive alcohol consumption and cancer; and between industrial and vehicular accidents and absenteeism. However, many employee-assistance programs which counsel troubled employees have demonstrated effectiveness.

Even less evidence is available on behavior modification or life-style programs, such as physical fitness or nutrition programs, where relationships between behavior and health status are not clearly defined.

Employer-based "wellness" and health-care programs are not a new phenomenon, although in the past a number of these programs were limited to executives or to employees with substance abuse and emotional problems, or were directed at hazardous occupational and environmental conditions. In the late 1880s and early 1900s, there were also comprehensive health and medical programs provided by employers, generally in areas of the country where medical-care services were not readily available. Most notable were the clinics and hospitals run by the railroad and mining industries.

The discussion that follows includes the stated rationale for the programs, an overview of the types of "wellness" programs, a partial listing of employers with such programs and the types of projects they provide, a review of the evaluation efforts to date and suggestions of programs worth pursuing, and some issues that employers and unions should consider when contemplating such programs.

Rationale for the Programs

The reasons for starting a wellness program are diverse and often related to the locus within the company that stimulated the program. There can be multiple rationales within the same company and hence different definitions of success. The most common reasons for developing wellness programs include:

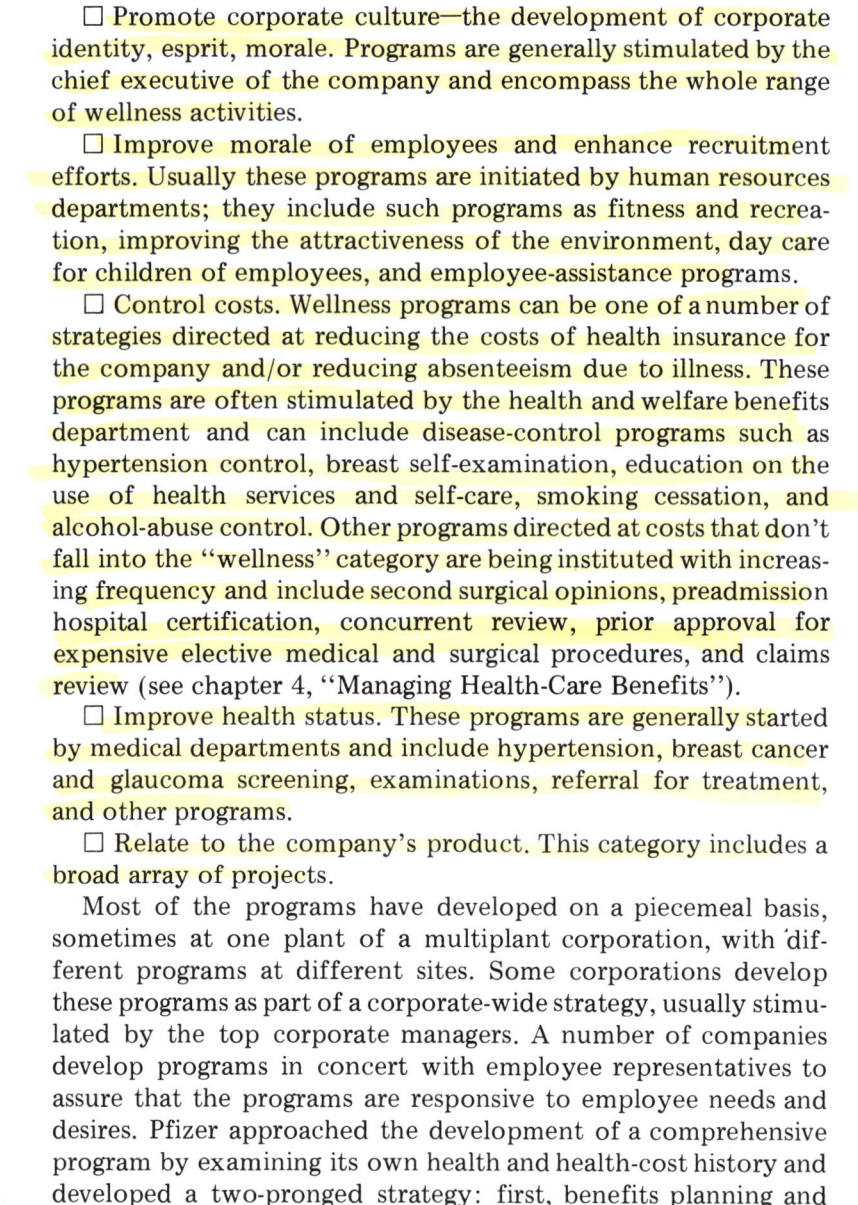

☐ Promote corporate culture—the development of corporate identity, esprit, morale. Programs are generally stimulated by the chief executive of the company and encompass the whole range of wellness activities.

☐ Improve morale of employees and enhance recruitment efforts. Usually these programs are initiated by human resources departments; they include such programs as fitness and recreation, improving the attractiveness of the environment, day care for children of employees, and employee-assistance programs.

☐ Control costs. Wellness programs can be one of a number of strategies directed at reducing the costs of health insurance for the company and/or reducing absenteeism due to illness. These programs are often stimulated by the health and welfare benefits department and can include disease-control programs such as hypertension control, breast self-examination, education on the use of health services and self-care, smoking cessation, and alcohol-abuse control. Other programs directed at costs that don't fall into the "wellness" category are being instituted with increasing frequency and include second surgical opinions, preadmission hospital certification, concurrent review, prior approval for expensive elective medical and surgical procedures, and claims review (see chapter 4, "Managing Health-Care Benefits").

☐ Improve health status. These programs are generally started by medical departments and include hypertension, breast cancer and glaucoma screening, examinations, referral for treatment, and other programs.

☐ Relate to the company's product. This category includes a broad array of projects.

Most of the programs have developed on a piecemeal basis, sometimes at one plant of a multiplant corporation, with different programs at different sites. Some corporations develop these programs as part of a corporate-wide strategy, usually stimulated by the top corporate managers. A number of companies develop programs in concert with employee representatives to assure that the programs are responsive to employee needs and desires. Pfizer approached the development of a comprehensive program by examining its own health and health-cost history and developed a two-pronged strategy: first, benefits planning and administration, which focused on structure content and payment systems; second, health-care cost management at the work site, which focused on injury and illness prevention, disability management, health promotion, and employee and family assistance. Efforts are being made by a number of companies with programs in existence to develop integrated strategies.

Classification of Programs

Wellness programs can be classified into four somewhat overlapping categories.

Early Intervention. Identification and treatment of disease or biological risks that lead to illness, including early-detection programs and disease-control programs. Early-detection programs incorporate screening for a number of health conditions, such as hypertension, diabetes, glaucoma, obesity, elevated cholesterol, and cancer. Periodic physicals are another variation on disease identification. Control programs usually incorporate referrals to physicians for treatment and/or the provision of direct medical care at the work site.

A number of programs that develop independently, such as alcohol or drug abuse and mental-health counseling, have been increasingly clustered under an umbrella called employee-assistance programs (EAPs). These rapidly growing programs, which offer prevention and early intervention, encompass more than health objectives.

The principal mission of EAPs is to help employees deal with a wide range of family and personal problems that affect, or can affect, job performance.

These problems have certain things in common:
—They are not "medical" in the customary sense, but are likely to have serious medical consequences if not resolved.
—Employees are unwilling to reveal them to the employer and/or fellow employees.
—Employees are unwilling to acknowledge some of these even to themselves.
—Some are caused or aggravated by stress on the job.
—The normal treatment for them is counseling and/or a change of life-style.

Because of these characteristics, many EAPs operate through links with community-based organizations, although they make it possible for the employee to contact hospitals or physicians.

Referral or direct services are provided for alcohol and substance abuse, mental-health problems, legal and financial problems, and marital problems. Supervisors can refer employees with problems, and employees in some firms can self-refer. Some programs have mandatory aspects. Joint union-employer programs of alcoholism rehabilitation have been especially successful.

Behavior Risks. Detection and control of high-risk behavior. Detection includes identification of smoking, poor nutrition, inactivity, stress, abnormal weight. Often risk-assessment profiles are used to detect these risks, or health fairs are sponsored by the employer. Control of high-risk behavior includes information, education, and counseling. Incentives are sometimes provided to encourage changes in behavior. Courses are sponsored or subsidized by employers, such as smoking cessation, weight reduction, physical fitness, and so on. Sometimes these programs are limited to people with the identified risks.

Healthy Behavior through Corporate Culture. These programs overlap the high-risk-control programs but are geared toward developing a corporate image and esprit and hence go beyond risk control. They include: company time for exercise and educational programs; flexitime, job sharing, maternity/paternity leaves; exercise and recreational facilities; nutritional food in cafeterias and vending machines; attractive work environments; and day care.

Reducing Exposure to Hazardous or Toxic Substances or Stress. While most of these programs would normally be classified as occupational health and safety, a few can be included in "wellness" programs. Such programs include: noise reduction, restricted smoking areas, improved ventilation, stress-reduction programs, required use of seat belts in company cars, information on home safety and accident prevention, and defensive driving courses.

Employers with Wellness Programs

A group of major corporations have been identified by the U.S. Department of Health and Human Services and others as providing health-promotion programs (see page 89). This list is not complete since hundreds of companies have some type of program. Although the Washington Business Group on Health has conducted several surveys, none has surveyed all types of employers.

Eligibility for programs varies widely and is not always corporate-wide. For example: IBM includes all employees, retirees, and spouses at all locations; Xerox, all employees at all locations;

Kimberly Clark, all employees, retirees, and spouses, but only at selected locations; Metropolitan Life, all employees, but only at selected locations; New York Telephone, high-risk employees; Texaco, executives only.

Examples of projects sponsored by major employers include:
- ☐ *Identification of disease, disease control.* Metropolitan Life, Baltimore Gas and Electric—hypertension control; Campbell Soup—colorectal cancer screening; IBM—multiphasic screening; Pioneer Hi Bred—screening with incentives.
- ☐ *High-risk behavior.* Coors, Hospital Corporation of America, City Federal Savings and Loan, Ford, Metropolitan Life, Campbell Soup.
- ☐ *Corporate culture.* Johnson & Johnson—a variety of programs; Hospital Corporation of America—fitness.

This short list is intended to provide a few examples of the diversity of programs. A growing number of employers sponsor comprehensive programs.

Evidence of the Effectiveness of Programs: Evaluation

There has been little systematic evaluation of these programs. Currently, Johnson & Johnson and Control Data have evaluation programs under way, and the Office of the Assistant Secretary for Planning and Evaluation of the U.S. Department of Health and Human Services is planning some additional evaluation programs. Most of the information up to now is anecdotal, or the information available is limited to employee satisfaction surveys. There are some early data from the Johnson & Johnson "Live for Life Program" that indicate behavior changes in participating employees. In other companies, except for a few isolated cost-benefit studies of hypertension control, cancer screening, and smoking cessation, and some evaluation of employee-assistance programs, few statistics have been collected.

The scarcity of studies is due to a number of reasons: Many of the programs emerge from the "good" idea of a manager and no evaluation is planned. Most programs do not identify objectives, nor state objectives in quantifiable terms—for example, 10 percent reduction in absenteeism, or lower hospitalization and medical-care costs for cardiovascular illness. In addition, many programs have multiple objectives.

A great many methodological and data problems have to be addressed before designing evaluation studies. Often, data needed for evaluation, such as medical utilization, sick-leave, and productivity measures, are not readily available. There are only a few quantifiable measures of change in health status that can be used in the short run. These include changes in diastolic and systolic blood pressure, blood-sugar levels, and serum cholesterol.

Longitudinal studies are necessary to trace most health-status changes. These studies are very expensive, and normal turnover in the workplace will shrink the sample over time. Some interventions, such as smoking cessation, weight loss, and alcoholism control, require tracking for recidivism for up to five years.

Utilization of medical-care services is subject to many influences that can confound a cause-and-effect relationship of an intervention. Such factors include ethnic differences in use of services, availability of services, recommendations of personal physicians, influenza epidemics, toxic exposure in the past, and genetics.

The great variety of programs require individualized evaluation studies and different measuring tools. For example, in measuring the success of disease identification and control programs, health-status measures are essential. Costs have to be measured in the short and long term. A physical fitness program, on the other hand, may not have measurable positive results for many years, but may incur short-term increases in absenteeism and medical bills for minor injuries.

Use of cost-benefit methodologies is difficult. The definitions of "costs" and "benefits" vary widely and agreed-upon measures of costs and benefits must be developed. All medical expenses must be captured. The discount rate used for determining loss of future earnings can cause large variations in results. The relationship of productivity measures to cost savings is still controversial. In addition to the foregoing problems, it is difficult to quantify "functional" changes and quality of life. Finally, in a relatively small firm, a few very large medical bills can distort the results.

For all these reasons, evaluation studies can be very complex and costly.

The literature suggests that companies until now have been more willing to publicize their successes than their failures. In addition, some of the scattered evaluations have been done by companies that market, or plan to market, wellness programs, which raises questions about the objectivity of the evaluations.

The literature suggests that expansion of wellness programs

has not been based on data that demonstrate a financial return to the corporation or changes in health outcomes. If, however, the goals of the program are an improvement in "corporate culture," "morale," or "recruitment," neither financial return nor health status is a relevant measure and perhaps programs designed for these reasons should not be classified as "wellness" or health programs.

There are few data on the costs of running the different programs, and these costs can vary widely for similar programs, depending on such factors as whether the employer provides the service directly, contracts on the premises for the program, contracts with providers of various types outside the workplace, or uses lay or professional staff to provide the service.

However, there are a number of evaluations or studies of programs which did not take place at the work site, and these can provide some guidance on efficacy for employers contemplating programs. For example, there is evidence that different types of hypertension-control programs will affect the success of reducing blood pressure levels and mortality. These studies show that programs that include periodic direct follow-up are more successful than programs that refer for treatment. There have also been evaluations of diabetes control and cancer screening and control programs.

National data are available on the relationship of heart disease, emphysema, and lung cancer to smoking, and it is clear that if smokers stop smoking, health outcomes improve. However, there is a lack of data on the efficacy of work-site smoking-cessation programs and differently designed programs. The data on effectiveness of non-work-site programs vary with the type of program. Also, there is a twofold variation on the costs of different smoking-cessation programs.

In addition, assigning costs of smoking over and beyond the cost of a cessation program is difficult. Annual *costs to employers* have been quoted as ranging from $336 to $601 per employee in the work force in one study. Another study assigns the *cost to society* at $190 per year.

Design of a program is important to its cost-benefit ratios. For example, if employers want to sponsor smoking-cessation programs, it is suggested by Jonathan Fielding, an expert on health promotion, that to maximize cost effectiveness, companies should pay for on-site cessation classes, contract only with programs that can validate a quit rate in a reasonable range, require copayment by employees to reinforce the commitment, and institute a clearly stated smoking policy that discourages smoking at the work site.

Evaluations of fitness programs suffer from numerous problems, such as self-selection by participants, inadequate controls, and lack of randomized studies; in addition, evaluations are skewed by the fact that employees' views of the health benefits derived from these programs are often based on their effect on morale rather than on health. Physical exercise, for example, has not yet been demonstrated to reduce cardiovascular incidents as many purport these programs will do.

Deciding Which Programs to Pursue

Before beginning programs or altering existing programs, it is necessary to assess the problems the programs will be designed to address, and to define in advance the objectives and expected outcomes. Once these steps are taken, an assessment is needed of the design of specific programs, their degree of success to date, and implementation plans. The following suggestions assume that the objectives are improvement of health status and/or reductions in health-care costs.

Programs with the Greatest Potential Payoff. The disease prevention and control programs are clearly those in which there is considerable evidence that intervention will make a difference. These include programs for hypertension, breast cancer, colorectal cancer, diabetes, glaucoma identification and treatment, and immunization. However, care needs to be exercised that mechanisms and support for follow-up are incorporated into screening programs.

Although occupational health and safety and disability management programs are not discussed in detail in this chapter, they can be included in prevention programs with considerable potential payoff. Such payoffs include avoidance of expensive medical bills, reduction in workers' compensation claims and disability benefits; reductions of absenteeism, tardiness, and turnover; and retention of productive employees.

Programs That May Have a High Payoff if Designed Appropriately. Smoking-cessation, alcohol-abuse programs, and employee-assistance programs may be useful if designed appropriately. Appropriate design, however, should include reinforcing features, such as discouraging smoking at the work site and/or alcohol at lunch, and paying for mental-health services that are recommended. Employee-assistance programs that have reinforcing features and pay for referral services can be quite successful in reducing absenteeism and increasing productivity. Often they prevent the loss of a productive employee and the destruction of a family.

Programs with Little or No Cost That May Affect Behavior. There are certain programs with minimal costs of implementation that can have some effect on behavior:
—Courses in use of health services
—Courses for specific health problems
—Requiring seat-belt use in company cars
—Changes in cafeteria food to include low-calorie, low-cholesterol, and low-sodium food, salads, and fresh fruit
—Company sponsorship of off-site team sports, such as bowling teams, marathons, bike trips, and so on
—Bike stands

Programs with Controversial or Questionable Value from a Health and Cost Perspective. There are a number of programs where there is little evidence of cost effectiveness or where there is evidence of negative cost-benefit relationships.
—Annual physical checkups and multiphasic screening without follow-up
—Risk-assessment programs without follow-up
—Physical-fitness programs requiring large company investments in membership fees or physical facilities
—General health-education courses
—Contracts with "wellness" or "fitness" firms for wellness programs, when the firm cannot provide verifiable evidence of success rates related to objectives.

Criteria for a Program

Given the lack of data on many of the programs and on their effectiveness and costs, selectivity and careful planning should precede the implementation of a program. The following is a suggested list of criteria which could be used to assess the advisability of planning a program, and steps that can be taken during the planning process.

1. *Defining the objectives.* The first step is to determine whether there are specific problems that need to be addressed

or whether the company is interested in a general program of health promotion. Objectives should be tied to specific problems that can be identified from company data, for example, age, race, occupation, utilization of medical care by employees, absenteeism, workers' compensation claims. For example, if the work force has a large number of black employees, hypertension screening and control might be a very important program to pursue, given the susceptibility of blacks to hypertension.

If the objective is to improve health status or to reduce medical-care premium cost, the package of programs might be very different from those designed only to control accidents or to effect a change in corporate culture.

2. *Assessing the potential needs of employees and the use of the program by employees.* Consultation with union representatives, other employee groups, and managers for their perception of problems and interests, and involvement by employees and unions in the planning process.

3. *Determining eligibility in advance.* Extending all programs to all employees may not be cost-effective, depending on the problem and the objective. Defining discrete groups eligible for a program can help control costs of expensive programs and/or target programs to risk groups where there will be a high payoff. Some firms start with a health-risk appraisal, or identification of a localized productivity or absentee problem, and focus the program on employees at risk.

4. *Assessing the capacity to provide the program, including the cost to the firm.* A company could directly provide the services or contract for on-site or off-site programs. Availability of space and personnel, community resources, commercial providers, and comparative costs of options need to be assessed. Some programs can be developed where employees themselves direct the program and provide peer support.

5. *Assessing current company practices that may be incompatible with the program.* If the cafeteria food does not offer choices of salads and low-sodium, low-sugar, and low-fat food, or if all vending machines dispense soda, coffee, candy, and cookies, these factors may be counterproductive to a nutrition education program.

6. *Trying to quantify the benefits versus the costs.* Although it is difficult, some data might be obtained from companies that have similar programs and from community-based programs. For

large firms, statisticians can make projections of expected reductions in risk and medical costs for certain types of programs.

7. *Trying the program on a demonstration basis.* In large multiplant companies, a demonstration with carefully designed evaluation could be a first step, with modification and/or extension of the program based on the evaluation. A number of companies with successful programs have started with demonstrations.

8. *Contracting for services.* A great many firms, including hospitals, academic health centers, and corporations, now offer health-promotion, wellness, and physical-fitness programs. Costs and outcomes vary widely. Purchasing these services should follow the company's prudent-buyer practices. The costs of providing the service in house as opposed to providing it through outside agencies should be compared with information obtained from employers who have used the agencies.

9. *Corporate commitment.* Programs can succeed or fail based on the commitment of corporate executives. This commitment can take many forms, such as participation by managers and incentives of various sorts, both monetary and nonmonetary.

10. *Confidentiality.* For disease control, alcohol and drug abuse, employee-assistance programs, medical evaluation, and risk assessment, advance plans have to be made and maintained to assure confidentiality of medical and personal information. If confidentiality is breached, the future usefulness of any program is questionable.

11. *Designing an evaluation plan.* These plans must be designed at the beginning of the program, with clear statements of objectives to be achieved, measurements to be used, and predefined indicators of failure or success.

Issues

There are a number of sensitive issues that need to be carefully addressed in designing and implementing wellness programs.

Confidentiality. Although there is no evidence of breaches of confidentiality, there are concerns about the uses that may be made of data from wellness programs, particularly risk-assessment and other programs that collect medical and personal information. Consequences of breaches of confidentiality can affect employees' future employment, insurability and insurance rates and, obviously, the future interest in, and success of, wellness programs.

Coercion/Incentives. Again, there is no evidence that coercion has occurred (except for the potential loss of employment in mandatory employee-assistance programs, where the employee refuses to cooperate). Many companies have used incentives to encourage participation in wellness programs. However, given rising health-insurance premiums, the temptation to increase employee participation in programs in order to reduce health-care costs could lead to subtle and not so subtle coercion.

Liability Risks. Increased risk of liability could occur with certain physical-fitness programs on the employer's premises, or there could be increased malpractice premiums for screening and treatment programs.

Competition, Quality Control, and Cost. The explosion of "wellness" firms and the scarcity of evaluation data will require more careful shopping for programs with track records and reasonable cost-benefit ratios. Three other factors affect this: the trend to emulate health-service delivery through credentials, licensure, and requiring graduate degrees for personnel; physician entrance into the market and efforts to control the market; movement toward proprietary firms and fee-for-service reimbursement. Until recently, off-site programs were often provided by nonprofit organizations like Ys, and health associations like the American Cancer Society. An increasing number of hospitals and commercial firms are entering the field, and the individuals providing the services are beginning to form exclusive professional groups.

Summary and Recommendations

Wellness is not a fad. Well-chosen, well-thought-out, well-executed programs will continue to grow. Others may fall by the wayside. Wellness programs, which are not a new phenomenon, appear to be spreading rapidly, particularly among large corporations. Often the programs spring out of different divisions of the company, with little careful or formal planning or evaluation protocols. Often the program's goals are not clearly stated; they may range widely from efforts to minimize and control known health risks through known interventions, to programs with conjectural benefits. The programs range from low- or no-cost interventions, such as vending machines with health foods, to programs requiring large investments, recruitment of medical personnel, or payment for specified medical interventions.

WELLNESS IN THE WORKPLACE: COMPANIES AND PROGRAMS

Listed below are a few of the companies that provide wellness programs for their employees. Many of these firms are recognizable as "big" companies but are comprised of many small plants scattered throughout the United States, often with fewer than 200 employees. Most of them have comprehensive wellness and life-style programs, covering a wide variety of topics; particularly noteworthy programs are highlighted.

Adolph Coors Company
American Hospital Association
American Hospital Supply Corporation
American Telephone and Telegraph
Anheuser-Busch Companies, Inc.
 *employee-assistance programs
Atlantic Richfield
 *employee-assistance programs
Blue Cross/Blue Shield of Indiana
Bonne Bell Cosmetics
Burlington Industries
Campbell Soup Company
 *early disease detection, screenings
Cannon Mills
 *early disease detection, screenings
Control Data Corporation
E.I. duPont de Nemours & Company
 *employee-assistance programs
General Dynamics
General Mills
General Motors
 *employee-assistance programs
General Telephone of Florida
 *physical fitness
Hospital Corporation of America
 *physical fitness

International Business Machines
Intermatic, Inc.
 *smoking cessation
Internorth, Inc.
 *physical fitness
International Telephone and Telegraph (ITT)
 *employee-assistance programs
Johnson & Johnson
Kimberly-Clark
Metropolitan Life Insurance Company
Pepsico
Phillips Petroleum
 *physical fitness
Pillsbury Company
Prudential Insurance Company
 *physical fitness
Sentry Life Insurance
Speedcall Corporation
 *smoking cessation
Tenneco
United Store Workers of America Health and Welfare Fund (Gimbels and Bloomingdale's)
 *high blood pressure control
United Technologies
 *employee-assistance programs
Xerox

Source: U.S. Department of Health and Human Services

The expansion of these programs, by and large, has not been based on data on effectiveness, financial return or positive cost-benefit outcomes to employers and employees, or on health status. Evaluation of the different types of programs is past due.

Great caution should be exercised before employers and employees begin extensive programs; careful planning, needs assessment, cost and benefit estimation, implementation plans, and evaluation plans should certainly precede their start.

RECOMMENDATIONS

12 Prevention is potentially one of the most cost-effective approaches to health-care management. Employers should survey their employees, dependents, and retirees, and determine which current and foreseeable health problems are most in need of attention. Based on the outcome of the survey, they should decide whether proven preventive methods of ameliorating these problems are available, and how cost-effective these methods would be in the circumstances of the given organization. To ensure that the work force gives full support to preventive measures, employers should involve employees and their union or unions in designing and carrying out the survey and any programs that flow from it.

13 When a survey discloses that certain preventive measures would be cost-effective, the employer and its union or unions should give top priority to those which have proved most successful in national or local trials, especially if the risk group is of appreciable size—for example, disease-prevention programs, and programs to screen and apply proven interventions to early stages of disease. Second priority should go to programs in which, if intervention succeeds, health outcomes could be improved and the cost of medical care reduced—for example, smoking cessation, substance moderation, mental health. Third priority should go to programs with sound objectives but still unproven efficacy—for example, physical fitness, nutrition/obesity control. Low-cost programs which may or may not have an impact on health status or cost of health care but which could improve morale and company esprit should also be considered—for example, attractive surroundings, health courses of various types, recreational activities, changes in cafeterias.

14 Employers and unions should enter into a preventive program only after developing clearly stated objectives, cost-benefit estimates, and plans for concurrent evaluation.

15 Employers and unions should ensure that preventive programs are reinforced by appropriate corporate practices, for example, smoke-free areas, mandatory use of seat belts in company cars, more nutritious cafeteria menus, reimbursement for prescribed mental-health services. They should also ensure that individual employees' health risks and problems will be kept in confidence.

REFERENCES

Allen, Richard. "Health Care Cost Management at the Worksite—the Pfizer Approach." An unpublished paper commissioned by Work in America Institute, January 31, 1984.

Curtis, Wilbur. "Live for Life: An Epidemiologic Evaluation of a Comprehensive Health Promotion Program." An unpublished paper from Johnson & Johnson.

Falik, Marilyn. U.S. Department of Health and Human Services. Discussion with author.

Fielding, Jonathan E. "Effectiveness of Employee Health Improvement Programs." *Journal of Occupational Medicine* 24 (November 1982).

Fielding, Jonathan E., and Breslow, Lester. "Worksite Hypertension Programs: Results of a Survey of 424 California Employers." *Public Health Reports* 98 (March-April 1983).

Kiefhaber, Ann. Washington Business Group on Health. Discussion with author.

Kiefhaber, Ann, and Goldbeck, Willis B. "Worksite Wellness." In *Health Care Cost Management*, edited by Peter Fox, Willis B. Goldbeck, and Jacob J. Spier. Ann Arbor, Mich.: Health Administration Press, 1984.

Kristein, Marvin M. "Economics of Health Promotion at the Work Site." *Health Education Quarterly*. Special supplement, 1982.

U.S. Department of Health and Human Services, Office of the Assistant Secretary for Planning and Evaluation. *Synthesis of Private Sector Health Care Initiatives.* Washington, D.C.: U.S. Department of Health and Human Services, 1984.

U.S. Department of Health and Human Services, Public Health Service. *Worksite Health Promotion: Some Questions and Answers to Help You Get Started.* Washington, D.C.: U.S. Department of Health and Human Services, 1983.

6. HEALTH MAINTENANCE ORGANIZATIONS

When health maintenance organizations (HMOs) first appeared, they were regarded as ugly ducklings by physicians, many corporations, and some of the general public. Today, they are seen as swans by almost all groups.

The concept is certainly valid, and much of the praise is well deserved, but excessive adulation should not be allowed to obscure the fact that HMOs differ markedly among themselves with respect to efficiency, quality of care, and fiscal soundness.

HMOs are on everyone's lips because they give health-care providers a clear *incentive* to dispense medical and hospital services efficiently. Their primary economic attraction, however, is that they reduce quite sharply the incidence of hospitalization, the largest single ingredient of health-care costs. But it is important to remember that HMOs are not the *only* mechanism for reducing hospitalization.

HMOs are a rapidly growing group of organizations that provide directly, or arrange for, comprehensive outpatient and inpatient services for a fixed payment per enrollee or family. They differ from traditional health insurance in that they manage an individual's or family's health care in addition to underwriting it. They differ from traditional fee-for-service medical practice in the following ways: First, the organization is "at risk" for providing the promised benefits at a capitation payment set in advance. Capitation payment serves as an incentive to control utilization of high-cost services. Traditional insurance, based on fee-for-service health care, pays providers for individual services; the more services providers supply, the more they earn. So the incentive is to increase units of service. Second, the HMO provides case management of the patients' health-care needs, including referral to specialists and choice of hospital or other provider. Patients cannot directly refer themselves for specialty care, hospitalization, or other services without incurring additional direct costs.

Health maintenance organizations have been on the scene for a long time. Ross Loos Clinic in Los Angeles is the oldest. One of the earliest and most successful HMOs was sponsored initially in 1938 by a major employer, Kaiser, to serve its own workers at a remote site in the State of Washington. This was the first venture by an employer to sponsor an HMO, although labor unions like the United Auto Workers tried to stimulate the development of these organizations. The Kaiser Plan later expanded to serve whole communities and, today, has branches across the country. However, for many decades, HMOs were slow to grow and were vigorously opposed by organized medicine. Opposition included efforts to deny membership to HMO physicians in local medical societies, to withhold hospital privileges, and to develop state legislation prohibiting the "corporate practice of medicine." In some sections of the country opposition continues. HMO growth was also impeded by the fact that most employment-based health-insurance plans did not offer HMO coverage as a choice. Employers, for the most part, were indifferent or negative.

The original and still dominant type of HMO was a multi-specialty group of physicians who practiced together at the same site. Physicians have developed an additional type of HMO, the individual practice association (IPA), in which the physician continues to practice in his or her own office.

The HMO concept is applicable to psychiatric and dental as well as physical health care. For example, the California Psychological Health Plan, a nonprofit group established in 1975, provides confidential psychiatric care to employees and their dependents at a set monthly fee per family.[1] Dental HMOs have also sprung up in many places.

In 1971, the Nixon Administration developed a white paper on health designed to address health-care costs and other health-care delivery problems. One of the major reforms proposed by the white paper was federal investment to stimulate HMOs, a term coined by Dr. Paul Ellwood. Legislation enacted in 1973 provided for the establishment of the federal HMO program, which offered grants and loans for the development of nonprofit HMOs. Not until 1975 did the federal government require employers to offer a choice of health plans, including an HMO, in addition to its standard insurance policy. This was critical to the expansion of HMOs because, previously, most employees could not elect this type of coverage and have the costs covered by the company health-insurance plan.

Until recently, the majority of HMOs were nonprofit, prepaid group-practice organizations, either community-based and -sponsored, like Group Health of Puget Sound and Group Health of Washington, or physician-sponsored, like Ross Loos and the Palo Alto Clinic. The individual practice association, which evolved in the 1970s, is frequently sponsored by physicians. Since 1981, there has been a rapid entry of for-profit HMOs and the development of multistate HMO companies.

HMOs have grown from 30 plans in 1970 to over 350 in 1985 and now serve more than 12 million people, about 5 percent of the population. Enrollment is projected to grow to 15 percent of the population by 1990.

What Is an HMO?

The term HMO—a relatively new nomenclature for what was once called a prepaid group practice—now covers a variety of organizational arrangements.[2] There are two basic forms, and each form has numerous variations:

Individual Practice Association (IPA). In this model the HMO contracts with an IPA, which contracts with physicians or physicians' groups on a capitation basis. Physicians who are part of these arrangements see HMO and non-HMO patients in their individual offices, and the physician's services are paid for on a fee-for-service basis. The primary physician is responsible for making referrals. The patient must use a referring physician who is under contract to the IPA. The physician bills the IPA on a fee-for-service basis, although the IPA is paid on a capitation basis. However, the physicians share the risk. If the estimated costs are exceeded, the physicians take a reduction in their individual payments.

Prepaid Group Practice (PGP). Under the prepaid-group-practice model, physicians are paid on either a salary or a capitation basis. They may be employed directly by the HMO or serve as members of a medical partnership which contracts with the HMO. The physicians practice together in a group. While some highly specialized services may be provided by outside physicians, in HMOs with large enrollments the majority of services are provided by the group. A few PGPs, most notably Kaiser, also own their own hospitals, but usually the PGP contracts with one or more hospitals.

Why Were HMOs Slow to Grow?

Many complex factors, including some of the following, have impeded the development of HMOs. There was strong early opposition by organized medicine, which still exists in certain areas of the country. Organized medicine traditionally has opposed salaried practice, institutional employment, and any deviation from fee-for-service payment. In addition, as recently as five years ago, it was virtually impossible to recruit physicians in high-paying specialties to accept a salaried position with a PGP. With the increased supply of physicians and the growing competition, this is no longer a major problem. The market for physician services has changed dramatically. In the past, HMOs also used to have a high turnover of physicians who sought HMO practice when they completed their residency to assure an income and then left as they accumulated resources to start a fee-for-service practice. Some HMOs were also heavily dependent on foreign-trained physicians to provide services. U.S.-trained physicians are now increasingly attracted to HMO practice.

With a few exceptions, it was very difficult for many consumers to subscribe to an HMO. Employers generally offered one insurance plan and would not pay for HMO membership as an alternative. Insurance companies did not underwrite these plans in the past, nor offer them as part of a package. However, in the last decade, a number of insurance companies have sponsored or established HMOs: Metropolitan Life, Prudential, Blue Cross/Blue Shield, CIGNA, and John Hancock. Although employer attitudes have changed dramatically, largely due to the requirement in the 1975 legislation, a majority of employers still do not actively promote HMO enrollment.

Employers, until recently, kept a hands-off policy on issues of delivery of health care. It is only in the last several years, as costs have continued to rise, that employers have sought to change their roles from payers of insurance to prudent purchasers of services.

The Health Care Financing Administration and the congressional committees responsible for Medicare, even after the 1973 amendments, were reluctant to enter into capitation arrangements with HMOs for Medicare beneficiaries. The federal government now actively stimulates enrollment of Medicare beneficiaries in HMOs.

The experience of the California Medicaid program with for-profit HMOs that were formed to care for Medicaid patients discouraged active support of HMOs. There was widespread fraud, abuse, and underutilization of services in these so-called HMO arrangements.

Some of the HMOs that developed at the beginning of the federal program were not financially or managerially viable and failed, leaving consumers and employers with problems of coverage and debts. Currently, in order to qualify to serve Medicare beneficiaries, HMOs must meet certain federal standards. The early experiences with HMOs have led to more prudent selection by employers and consumers, and one of the criteria they use is whether the HMO is federally qualified. However, an investigation of HMOs that were developed to serve Medicare patients in Florida is currently being conducted by the General Accounting Office of the U.S. Congress in response to charges of poor service.

Consumers who have established relationships with primary-care physicians and are satisfied with their care are reluctant to switch to HMO coverage. Furthermore, under many company plans, HMO enrollment requires a greater out-of-pocket premium payment by the employee than a standard insurance policy. However, there are usually no deductibles or coinsurance under HMO plans, and HMO services are often more comprehensive than standard insurance policies.

A problem remains in developing and managing HMOs. Up to five years may be required before PGPs reach the stability and economies of scale needed for success. The larger the HMO enrollment, the more feasible it is to provide most specialized services directly rather than through contract. Both front-end financing for nonprofit HMOs and recruitment of trained managers have been difficult.

There are also some subtle staffing problems that have impeded, and may continue to impede, growth. There has been a lack of training of medical students for HMO practice and for work with other health professionals as a team. HMOs have been reluctant to participate in health-professional training programs because of the cost, although some HMOs, like Kaiser or university-sponsored HMOs, provide training opportunities. In addition, the faculty at some medical schools are reluctant to supervise training at HMOs because they are outside of the hospital-medical school base. Some difficulty has also been encountered in recruiting physicians because they have much less autonomy in an HMO setting.

In the spring of 1979, the federal Office of Health Maintenance Organizations conducted a study which isolated factors that had slowed the growth and development of HMOs in the Detroit area, the headquarters of the major auto companies and the United Auto Workers.[3] There were six HMOs in the Detroit area at that time, but enrollment was growing very slowly despite the UAW's long-time support of HMOs.

Findings of the study were:

☐ Total HMO premiums and benefits were competitive for comprehensive service, but there was general satisfaction with the existing Blue Cross plans.
☐ There was little financial advantage for employees in enrolling in an HMO. Many benefit plans were noncontributory and pegged at traditional premiums. If the HMO costs were higher (because more comprehensive benefits were offered), the employee paid the difference. If they were lower, the employer kept the savings.
☐ Employers were less cooperative than in more active HMO areas (this factor has changed).
☐ Most labor unions other than the UAW were not supportive. The marketing effort addressed to unions by the HMOs was inadequate.
☐ Physicians in the area were not supportive.
☐ Suburban hospitals were not cooperating with HMOs.
☐ Blue Cross/Blue Shield had started an IPA that competed with the existing PGPs.

How Well Do HMOs Perform?

Two recent publications provide information on HMO performance: the Rand study and a study conducted in 1981 by Harold Luft of the University of California, San Francisco.[4] The Luft study produced a synthesis of literature and research on the performance of HMOs for the National Center for Health Services Research. The study found that there are considerable differences in performance between the PGP and IPA HMO models. For example:

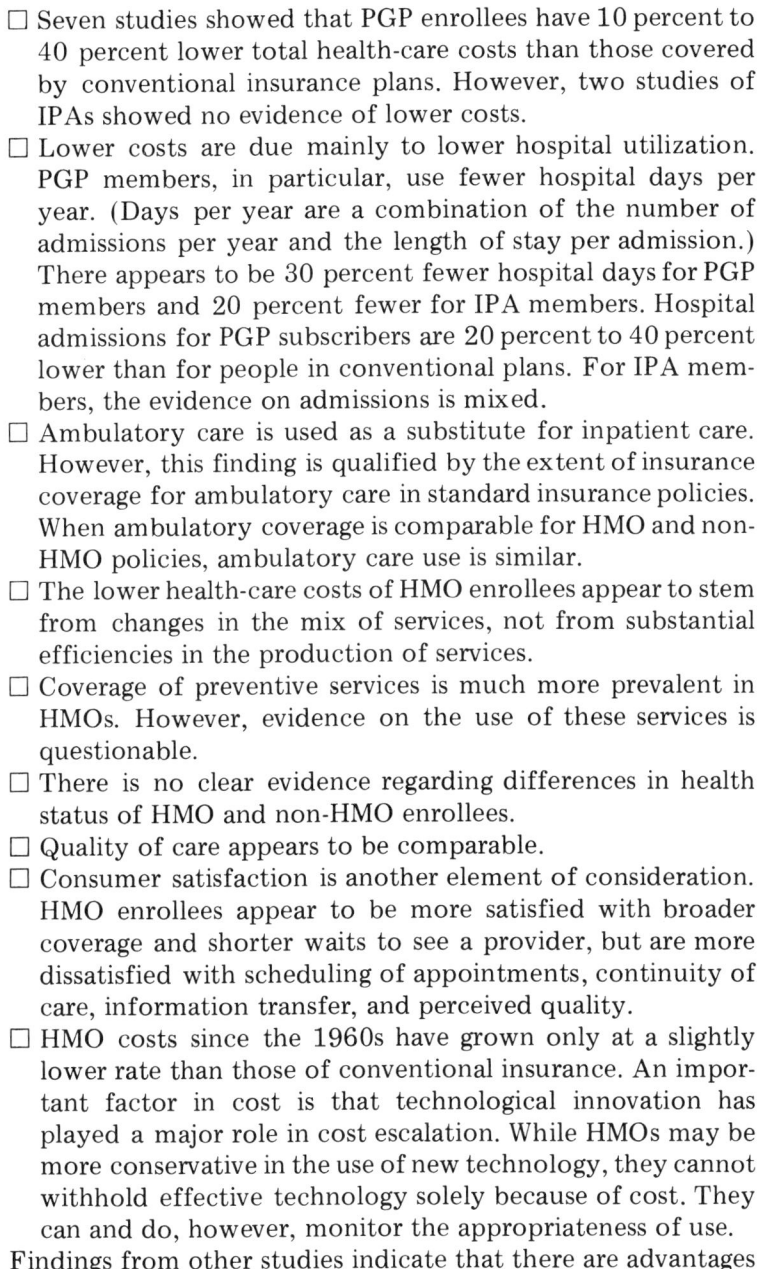

- Seven studies showed that PGP enrollees have 10 percent to 40 percent lower total health-care costs than those covered by conventional insurance plans. However, two studies of IPAs showed no evidence of lower costs.
- Lower costs are due mainly to lower hospital utilization. PGP members, in particular, use fewer hospital days per year. (Days per year are a combination of the number of admissions per year and the length of stay per admission.) There appears to be 30 percent fewer hospital days for PGP members and 20 percent fewer for IPA members. Hospital admissions for PGP subscribers are 20 percent to 40 percent lower than for people in conventional plans. For IPA members, the evidence on admissions is mixed.
- Ambulatory care is used as a substitute for inpatient care. However, this finding is qualified by the extent of insurance coverage for ambulatory care in standard insurance policies. When ambulatory coverage is comparable for HMO and non-HMO policies, ambulatory care use is similar.
- The lower health-care costs of HMO enrollees appear to stem from changes in the mix of services, not from substantial efficiencies in the production of services.
- Coverage of preventive services is much more prevalent in HMOs. However, evidence on the use of these services is questionable.
- There is no clear evidence regarding differences in health status of HMO and non-HMO enrollees.
- Quality of care appears to be comparable.
- Consumer satisfaction is another element of consideration. HMO enrollees appear to be more satisfied with broader coverage and shorter waits to see a provider, but are more dissatisfied with scheduling of appointments, continuity of care, information transfer, and perceived quality.
- HMO costs since the 1960s have grown only at a slightly lower rate than those of conventional insurance. An important factor in cost is that technological innovation has played a major role in cost escalation. While HMOs may be more conservative in the use of new technology, they cannot withhold effective technology solely because of cost. They can and do, however, monitor the appropriateness of use.

Findings from other studies indicate that there are advantages to PGPs over IPAs as follows:

☐ PGPs show greater cost savings, particularly large PGPs.
☐ PGPs tend to stimulate greater competitive response from traditional providers than do IPAs.[5]

The most recent study of HMO experience was conducted by the Rand Corporation as part of a large health-insurance experiment funded by the federal government. The analysis of health maintenance organizations compared individuals and families who had been randomly assigned to receive care at an HMO with no cost sharing with those who received care in the fee-for-service system with no cost sharing.

The report concluded that the estimated value of services delivered to the HMO group was 28 percent less than the value of services delivered to the comparable fee-for-service group. The principal cause of the difference was in the hospital admission rate, which was 40 percent less in the HMO group than in the fee-for-service group.

Although some observers argue that the HMO's emphasis on prevention is what accounts for its enrollees being hospitalized less often than people on fee-for-service plans, that theory does not hold up. When the Rand Corporation compared the behavior of people on various fee-for-service plans, it found the opposite relationship; that is, the fee-for-service groups who made the greatest number of preventive visits were hospitalized more often than the rest.

The markedly lower rate of hospitalization at HMOs appears to reflect a different style of medicine. An analysis of data on health status and patient satisfaction at the HMO, compared with fee-for-service, is in progress.

There are other factors not discussed in the cited studies that may have a bearing on the experience with the PGP form of HMOs. For example, the small-area studies of physician practice variation by Dr. John Wennberg of the Dartmouth Medical School and the community' style of medical practice are probably relevant to HMOs: The PGP style tends to be more conservative in regard to utilization of services, particularly elective surgery and hospitalization.

When physicians practice together in a group setting and see one another's patients, there is greater opportunity for peer review, peer consultation, and "corridor consultation." There may be less uncertainty in diagnosis and treatment decisions. As a result, there may be less need for hospital admissions, extra tests, callbacks, and longer hospital stays. In addition, physicians in a PGP receive timely, organized feedback on expenses incurred per patient, by physician and diagnosis.

There are economies of scale in a PGP and the opportunity to use health professionals other than physicians to perform functions within their skills at lower costs. Prepaid groups have made greater use of nurse practitioners, physicians' assistants, and social workers than have traditional practitioners.

There may be a certain amount of self-selection, with consumers who do not like waiting or receiving care from different physicians opting out of HMOs. The Rand study, although controlling for self-selection, excluded the disabled, institutionalized population; end-stage renal-disease patients; and those over 62. Many critics have argued that HMOs enjoy lower hospital utilization because fewer older people are enrolled in them. However, a recent survey proves the contrary. It is true that older HMO enrollees use more days of hospital care than younger enrollees do; but, when over-65 HMO enrollees were compared with over-65 non-HMO enrollees, the former used 2,084 days per thousand, as against 4,382 days per thousand for the latter.[6]

In addition, HMOs tend to enroll groups rather than individuals and to market to larger employer-employee groups, which may also affect the mix of patients. There are also consumers who seek medical care for more than strictly medical reasons—"the worried well." While some studies have been done on the distribution between HMOs and non-HMOs based on population differences and health-status differences, they have not been conducted on the social factors that influence choice.

The rapid growth of HMOs and the entry of for-profit HMOs are relatively new, and there are no studies that provide data on cost effectiveness, morbidity, or patient mix between for-profit and not-for-profit HMOs.

Can the Competition of HMOs Influence Other Providers and Contain Costs?

There is still too little experience to reach definitive conclusions on the issue of whether HMOs introduce sufficient competition to influence other providers and contain total costs. Only in California; Minneapolis; Rochester, New York; and, perhaps, the quad-city area of Illinois and Iowa (Moline and Rock Island, Illinois, and Davenport and Bettendorf, Iowa), is the balance between HMOs and non-HMOs great enough to induce competition of this kind.

In California, total medical-care costs remain among the highest in the nation, and the rate of escalation appears to be similar to the national rate. In Minneapolis and Rochester, on the other hand, there appears to have been a moderating effect. In Rochester, however, there are two confounding factors: (1) hospital-rate regulation, and (2) a long tradition of community interest and involvement in health issues, with the active participation of the two dominant employers in the area, Kodak and Xerox. In Minneapolis there are also active employer-sponsored programs of claims review and preadmission certification for non-HMO enrollees, and these could well influence the cost trends as much as the penetration of HMOs.

A number of commentators have noted that the *potential* of HMO entry into a market often stimulates providers of care to form other types of organizations, such as preferred provider organizations (PPOs), to compete. Providers become more receptive to negotiation with employers and unions.[7] Frequently physicians form an IPA when a PGP-HMO enters the area.

Employer-Union Involvement in HMOs

For many years, except for the Kaiser plan and the railroad hospital and clinic plans (the latter ended in the 1950s), employers played little or no role in the development or support of HMOs. In fact, as previously noted, refusal to offer choice impeded growth. Many companies still oppose mandating the dual-choice option.[8] For small employers, offering options can add considerable administrative expense to their health-benefits program. Small employers also do not have the resources for producing comparative brochures or assessing the viability of HMOs. Also, HMOs rarely seek out small employers for marketing.

Unions have played a far more active role in stimulating and encouraging HMO development and enrollment. Federal and state employee unions, the UAW, and other union groups were active in encouraging the development of HMOs and encouraging enrollment. Some unions, such as the International Ladies' Garment Workers' Union, actually ran (and still run) HMO-type clinics. Other unions, however, opposed HMOs as a reduction of their freedom of choice.

Recently a number of employers have begun to stimulate,

sponsor, and underwrite HMO development, and insurance companies that sell traditional health insurance policies have become deeply involved in the formation of HMOs.

The range of involvement varies, but includes the following:

—Stimulation of interest of employees to enroll in HMOs where they exist. Some companies welcome active marketing by HMOs.

—Identification and evaluation of HMOs to assure that those selected for health benefits plans meet quality and fiscal standards.

—Participation of executives on boards of HMOs.

—Sponsorship of feasibility studies and provision of front-end financing through loans to assist in the establishment of HMOs.

—Actual sponsorship or ownership of an HMO.

Examples of recent employer/union involvement include the following:

☐ Employers in the Minneapolis-St. Paul area stimulated development of HMOs, beginning in 1974, through the Twin Cities Development Project. In 1972, 2 percent of the population of the area received care from HMOs. Honeywell, the last major employer in the area to actively participate, now has enrolled 60 percent of its work force; General Mills has enrolled 70 percent.

☐ R.J. Reynolds and Deere and Company have set up their own HMOs.[9] Reynolds owns its HMO. Deere helped finance an HMO to serve Moline and Rock Island in Illinois and Bettendorf and Davenport in Iowa; it has also helped to establish a second HMO in Waterloo, Iowa.[10]

☐ Ford and the UAW helped establish an HMO in Detroit in 1978. The big three automakers brought Kaiser into Detroit and converted a shaky HMO into the successful Health Alliance.

☐ Alcoa in Pittsburgh invited Prudential Insurance Company and the Kaiser Foundation to do a feasibility study related to setting up competitive HMOs.

☐ Caterpillar and Hewlett-Packard encourage enrollment of employees. Caterpillar made a $1 million line of credit available to start an HMO. Hewlett-Packard, which had problems with HMO insolvencies, now assesses the viability of any HMO selected for employee enrollment.

□ Lockheed requires all new employees to enroll in an HMO for the first year of employment.

Current Employer Attitudes toward HMOs[11]

The Office of Health Maintenance Organizations surveyed employer attitudes toward HMOs in 1983. The findings were as follows:

□ Forty-six percent of employers were positive toward the HMO concept; 21 percent were positive toward promoting employee enrollment; 23 percent were willing to support the formation of new HMOs where none exists.

□ Less than 5 percent had a negative overall policy.

□ Positive activities of employers to stimulate employee enrollment included the following:
  —70 percent have employer or HMO-run sales meetings on company time.
  —75 percent allow visits by HMO representatives.
  —85 percent printed a brochure which compares HMO and other benefits.

□ Employers reported a high degree of satisfaction with the marketing and enrollment, administrative procedures, and medical services of HMOs.

□ Twenty percent responded that they would be interested in supporting the development of new HMOs; 12 percent were willing to sponsor an HMO; 7 percent were willing to provide financial support; and 65 percent would allow employees to serve on an HMO board of directors or development committee.

□ Employers did not report too many problems with HMOs. There were a few, however.
  —Some reported inflexible procedures and difficulty in producing utilization reports.
  —14 percent reported higher HMO administrative costs.
  —10 percent experienced additional administrative expenses from HMO insolvency; 5 percent, additional premium expenses; and 4 percent, takeover of liability of outstanding claims.

□ Employers reported the following pros and cons from the employee perspective:

Pros
—Minimum out-of-pocket costs

—Comprehensive coverage
—Preventive services

Cons
—Reluctance to change physicians
—Limitation on choice of physicians, hospitals, and other providers

Employers and unions have also complained about the lack of data provided by HMOs on utilization and detailed costs, and also about the setting of premiums just below those of competitive plans so that savings are not shared with users.

Accessibility of HMOs[12]

In many parts of the country HMOs are still not available, or they have barely penetrated the market. For example, in 1983 only eight states had HMOs whose market shares exceeded 5 percent: Washington, California, Oregon, Arizona, Colorado, Minnesota, Wisconsin, and New York. Twenty-two states had no plans or less than five plans within the state: Idaho, Nevada, Montana, Wyoming, New Mexico, North Dakota, South Dakota, Nebraska, Kansas, Arkansas, Louisiana, Iowa, Illinois, Mississippi, Alabama, South Carolina, North Carolina, Virginia, West Virginia, New Hampshire, Vermont, and Maine.

Because the prepaid group-practice model of an HMO requires sizable enrollments for viability, areas that are sparsely populated and have low population densities will have difficulty establishing viable PGP-type HMOs, unless they are part of multiorganizational HMO systems.

It remains to be determined if HMOs will seek out smaller employers and individual enrollees. The question of adverse selection also remains unanswered. HMOs, by their nature and in their marketing, may result in adverse selection, with younger, more mobile, employed populations enrolling in these plans, while the disabled, those with chronic illness, and retirees continue in traditional plans. Also, if HMOs, clinics, and hospitals concentrate in suburban or more affluent urban neighborhoods rather than in poorer areas, lower-wage and blue-collar workers will not have ready access to these plans.

It is estimated that by 1990 HMOs will serve 15 percent of the population, a relatively small proportion to have a major competitive impact except in selected locations, where penetration is concentrated.

What Employers and Unions Can Do to Stimulate HMO Growth and Employee Enrollment

Successful efforts by employers to stimulate HMO growth can serve as examples for other employers who have the same objective.

Education, Information, and Incentives to Join HMOs. The following employers and unions have offered various incentives and undertaken extensive educational programs to interest employees in joining HMOs.

- *Ford Motor Company* launched an extensive education and communications program, including a personal letter from the board chairman, encouraging employees to explore the HMO alternative.
- *Chrysler* offered savings bonds up to $250 to its employees who were members of HMOs if they recruited other workers to join.
- *Deere*, as mentioned before, provided the impetus for HMO development and strongly encouraged enrollment. Forty percent of its employees are now enrolled.
- *IBM* offers 158 HMO plans across the country to its employees.
- The *United Auto Workers* and government employees' and teachers' unions have played major roles in encouraging employers to offer HMO options and to stimulate enrollment.
- The *Governor of New York* is seeking to double HMO enrollment of state employees by 1986. Active marketing campaigns have been planned by the state.

One of the most important roles for employers and unions is to select HMOs wisely for employees and monitor their performance. The same policies should be followed for other providers as well. This review should be ongoing, comprehensive, and include:

- Financial viability of the HMO. A number of HMOs have become insolvent in the last few years. With the recent rapid entry of for-profit HMOs, a pattern of abuse similar to that of California Medicaid experience or the nursing-home industry may emerge. Hewlett-Packard conducts extensive reviews of HMOs before offering them to employees.

☐ The quality and scope of coverage and arrangements with providers. Are the physicians qualified in their specialties? What are the reputations of the hospitals involved? What arrangements are made for provision of tertiary care and other highly specialized services, for emergency services, for out-of-area services? Are facilities and hours of service convenient?
☐ Utilization data and morbidity and mortality data.
☐ Employee satisfaction.
☐ Adverse selection. Are the total costs of health care for the employment unit changing, or is the HMO caring for only the lower-risk employees?

On the other hand, some employers still offer traditional plans with lower premiums and less comprehensive benefits and require employees to pay an extra premium for the HMO. Readjustments may be needed to balance the policies.

Some employers still make it difficult for employees to make wise choices by placing obstacles in the way of active HMO marketing and failing to provide a comparative analysis of choices, including total cost.

Standing up to Providers. In a number of communities, resistance by providers has made employers reluctant to fight for HMOs. However, large employers and unions have overcome provider opposition, once providers were convinced of the need to respond. Strong employer support can overcome provider opposition, particularly now, with the increased supply of physicians and declining hospital occupancy rates. Employers and unions should treat providers of health care no differently than providers of other products with respect to quality and price.

Bringing Management and Financial Expertise to Bear on the Development of HMOs. Successful HMOs have been started under the aegis of several corporations, which have provided:
—Feasibility studies and market analyses
—Organization of consortia of employers and unions who might participate
—Seed money for the development of the HMO and marketing expertise
—Recruitment of key HMO leadership
—Organizational and managerial expertise

The withdrawal of federal HMO support and the unavailability of start-up capital through grants and low-interest loans has made it increasingly difficult to start community-based HMOs.

Unanswered Questions about HMOs

There are a number of unanswered questions about HMOs and their future role. These include:

- [ ] Will the growing entry of for-profit HMOs and multistate HMOs lead to skimming, increased costs (because of profit margin expectation and return on equity), increased bankruptcies, and underutilization and inappropriate utilization of services, as marginal firms begin to enter the market?
- [ ] If penetration of HMOs reaches a third or half of the population in an area, will costs escalate at the same rate as for other forms of insurance and the cost savings disappear over time? Will HMOs enroll or attract only the lower-risk groups, forcing the higher-risk, higher-cost patients into more traditional care arrangements?
- [ ] Will HMOs actively market to smaller employers, or will premiums for small employers rise at a more rapid rate as large-employer groups gain access to HMOs? Will there be a growing differential in costs for insuring small groups and individuals and a negative reaction to HMOs from these groups?
- [ ] Will HMOs locate in areas where there is easy access for lower-income workers or locate mainly in suburbia and affluent urban sections? Will HMOs become the vehicle for care of only the higher income white-collar and blue-collar workers?
- [ ] Will the newly developed preferred provider organizations (PPOs) be competitive enough in the short run to slow the growth of the PGP form of HMOs?
- [ ] If increases in cost sharing become more common, will the difference in costs between HMOs and traditional coverage make HMO premiums comparatively higher and reduce employer enthusiasm for them? Will increased cost sharing lead to a reduction in the scope of benefits, reducing whatever effect expanded ambulatory-care services and preventive services have on hospital admissions?
- [ ] Will enough physicians be willing to work in HMO arrangements to sustain rapid growth?
- [ ] Will employers encourage retired employees to enroll in HMOs?
- [ ] Can consumer interest sustain increased growth? Is the demography of the population supportive of further rapid growth?

Lastly, bearing in mind that the key advantage of HMOs lies in their lower rates of hospitalization, there is a reasonable likelihood that fee-for-service physicians, stung by competition, may regain status by learning to resort less to inpatient services. Such techniques as greater use of ambulatory care, preadmission certification, utilization review, and the development of consensus on the management of cases where great variability now exists have been shown to cut hospitalization rates to a point where there remains little choice between fee-for-service health care and an HMO.

RECOMMENDATIONS

16 Employers and unions should encourage employees, dependents, and retirees to join HMOs that meet acceptable standards of cost, quality of care, fiscal soundness, and hospital utilization. It is now established that HMOs can furnish health care comparable to that provided by fee-for-service physicians, and at lower cost.

17 After evaluating an HMO and determining that it meets the standards, employers and unions should encourage enrollment by:
—removing financial and organizational disincentives, if any
—offering employees positive incentives to join
—informing employees about available choices among plans
—allowing the HMO to make sales presentations at the workplace

18 If existing HMOs in the area do not meet the standards, employers and unions should assist those that can be brought up to standard by reasonable means.

19 In areas where no HMOs exist, employers and unions should help to create them. They can do so by providing:
—managerial and marketing expertise
—grants for feasibility and marketing studies
—start-up loans

—help in organizing consortia with other employers and unions
—recruiting key leadership for incipient HMOs

NOTES

1. Victoria George and William E. Hembree, *Breakthroughs in Health-Care Management: Employer and Union Initiatives*, Pergamon Press/Work in America Institute Series (New York: Pergamon, 1986), chapter 4.
2. Harold Luft, *The Operations and Performance of Health Maintenance Organizations—A Synthesis of Findings from Health Services Research* (Rockville, Md.: National Center for Health Services Research, 1981).
3. U.S. Department of Health and Human Services, *Case Study Report on the Competitive Impact of HMOs in Detroit*, DHHS Publication No. (PHS) 81-50144 (Washington, D.C.: U.S. Department of Health and Human Services, 1981).
4. Luft, *The Operations and Performance of Health Maintenance Organizations*.
5. U.S. Department of Health, Education and Welfare, *National HMO Development Strategy through 1988, September 1979*, DHEW Publication No. (PHS) 79-50111 (Washington, D.C.: U.S. Department of Health, Education and Welfare, 1979).
6. Interstudy, *Hospital Utilization in HMOs*, HMO Industry Report Series, No. 5 (Excelsior, Minn.: Interstudy, 1984), as cited in James R. Kimmey, "Dealing with Excess Hospital Capacity: Methods and Consequences," an unpublished paper commissioned by Work in America Institute, September 16, 1984.
7. M. Reisler, "Business in Richmond Attacks Health Care Costs," *Harvard Business Review* 63 (January-February 1985): 145-155.
8. John K. Iglehart, "Health Care and American Business," *New England Journal of Medicine* 306 (January 14, 1982): 120-124.
9. U.S. Department of Health and Human Services, Office of Health Maintenance Organizations, *A Case Study of Industry for HMO Development—Quad City Health Plan*, DHHS Publication No. (PHS) 81-50160 (Washington, D.C.: U.S. Department of Health and Human Services, 1981).

10. John Perham, "New Funding for Health Care," *Dun's Business Month*, March 1983.
11. U.S. Department of Health and Human Services, Office of Health Maintenance Organizations, *Employer Attitudes toward Health Maintenance Organizations*, DHHS Publication No. HRSM-HM 83-2 (Washington, D.C.: U.S. Department of Health and Human Services, August 1983).
12. National Industry Council for HMO Development, *The Health Maintenance Organization Industry Ten Year Report, 1973-83* (Washington, D.C.: National Industry Council for HMO Development, U.S. Department of Health and Human Services, 1984.)

Also, discussions with Tony Masso and Frank Suebold of the Office of Health Maintenance Organizations, Department of Health and Human Services, and Willis Goldbeck, Washington Business Group on Health.

REFERENCES

Allen, Richard. "Health Care Cost Management at the Worksite— the Pfizer Approach." An unpublished paper commissioned by Work in America Institute, January 31, 1984.

Fox, Peter D; Goldbeck, Willis B.; and Spies, Jacob J. *Health Care Cost Management*, Ann Arbor, Mich.: Health Administration Press, 1984.

Fuchs, Victor R. "The Battle for Control of Health Care." *Health Affairs* (Summer 1982).

Hanft, Ruth S. "Alternatives to Hospital Care." An unpublished paper commissioned by Work in America Institute, May 9, 1984.

Hembree, William E. "Joint Labor and Management Efforts to Control Employee Health Care Costs." An unpublished paper commissioned by Work in America Institute, May 9, 1984.

Light, Donald W. "Is Competition Bad?" *New England Journal of Medicine* 309 (November 24, 1983).

Luft, Harold. *The Operations and Performance of Health Maintenance Organizations—A Synthesis of Findings from Health Services Research*. Rockville, Md.: National Center for Health Services Research, 1981.

Moxley, John H. III, and Roeder, Penelope. "New Opportunities for Out-of-Hospital Health Services." *New England Journal of Medicine* 310 (January 19, 1984).

Newhouse, Joseph P., et al. "Some Interim Results from a Controlled Trial of Cost Sharing in Health Insurance." *New England Journal of Medicine* 305 (December 17, 1981): 1501-1507.

Relman, Arnold S. "The New Medical Industrial Complex." *New England Journal of Medicine* 303 (October 23, 1980): 963-970.

Rice, Thomas H. "The Impact of Changing Medicare Reimbursement Rates on Physician-Induced Demand." *Medical Care* 21 (August 1983).

Saline, Lindon E. "An Overview of The Business Roundtable Health Initiatives." An unpublished paper commissioned by Work in America Institute, January 31, 1984.

7. *CORPORATE AND UNION STRATEGIES FOR HEALTH CARE*

The number of employers who are making concerted efforts to contain health-care costs has grown rapidly in the past few years, and individual companies are introducing more and more programs. For example, a survey of 1,185 companies in early 1984 found that 63 percent assessed deductibles for inpatient services, as against 30 percent in 1982; 47 percent were self-funded, as against 37 percent in 1982; 26 percent had preadmission requirements, as against 2 percent in 1982; about 70 percent were planning to conduct some form of claims review; about 28 percent had adopted one or another form of incentives to reduce hospitalization; and roughly 50 percent offered some kind of preventive and wellness program.[1] The recent perceptible slowing of health-care cost inflation is bound to reinforce this trend.

A few leading companies have recognized that it is necessary to organize their diverse activities into a coherent strategy, to ensure that programs are synergistic, or at least congruent. In some cases, major insurance companies have served as advisors. While experience is still too limited and tentative for confident recommendations, several examples may help to stimulate further useful experiment.

Common sense, bolstered by experience to date, suggests several ingredients for any comprehensive strategy:
 ☐ Examining the organization's health-care problems, forecasting what is likely to happen without intervention, and selecting some promising targets for improvement
 ☐ Surveying the data sources currently available to the organi-

zation and estimating what data will be needed in order to manage health care more successfully
☐ Setting practical goals for the improvement of health-care management
☐ Assessing and redesigning the health-benefits plan
☐ Developing methods for managing the health-benefits plan
☐ Obtaining active, visible support from top management
☐ Assigning responsibilities for each part of the strategy and providing staff, financial, and informational resources.

THE BUSINESS ROUNDTABLE[2]

In February 1982 The Business Roundtable gave a powerful and timely push to the concept of health-care management. In its pamphlet "An Appropriate Role for Corporations in Health-Care Cost Management," The Business Roundtable's Task Force on Health found that "the cost aspects of the health-care system warrant increased management attention," adding that "appropriate attention must continue to be given to questions of access to and quality of care." Further, it said, "Local action can be instrumental in slowing the rise in costs, but it is not appropriate to tinker unduly with systems that are among the best in the world. However, there are ample opportunities to share management skills and to show serious concern among leaders in the community-at-large. . . ."

The pamphlet pointed to three general areas "in which private-sector effectiveness has been demonstrated and application of additional leadership would be most beneficial to society as a whole":

☐ "Community involvement . . . on demonstration projects designed to stimulate competition . . . (and) in community coalitions established to address specific local health-care cost management problems."
☐ "Company programs . . . to improve employee health, to test innovative health-benefit packages and to continue the education of company officials who also serve as hospital trustees."
☐ "System checks . . . by participation in local health planning units and support of health utilization review committees. . . ."

Business Roundtable members were encouraged to "(1) reexamine the full spectrum of health-care cost issues at their company and community levels; (2) develop an action program responsive to their particular circumstances."

Members were urged to consider four broad areas of importance within the company:

- ☐ *Plan design.* In particular, consumer cost sharing to stimulate more prudent use of health services; second surgical opinions, preferably mandatory; reimbursement designed to favor less costly types of care; preadmission testing; ambulatory surgery; coverage for home health care; special incentives to discourage unnecessary use of resources.
- ☐ *Alternative financing arrangements.* For example, minimum premium plans, contracts for administrative services, self-insurance, prepaid per capita financing.
- ☐ *Claims control,* based on the use of reliable claims data, norms for identifying trouble spots, and effective provider relationships; special attention to conformance to plan provisions, appropriateness of treatment, appropriate fees, and utilization review.
- ☐ *Health promotion,* including a major commitment to health education, disease prevention, employee-assistance programs, and programs to modify unhealthy life-styles.

For action at the community level, The Business Roundtable advocated the following:

- ☐ *Cooperation with public institutions,* for example, employers and unions working jointly to increase the influence of health-systems agencies; contracting with Professional Standards Review Organizations (PSROs) for utilization reviews; encouragement of HMO membership and assistance in developing HMOs; actions to oppose the shifting by hospitals of unreimbursed Medicare/Medicaid costs to private users.
- ☐ *Business executive involvement* in Health Systems Agencies (HSAs), hospital boards, HMOs, and the like, and educating these executives to do a good job.
- ☐ *Local health-care coalitions* in cooperation with *all* interested groups.
- ☐ *Competition strategy,* that is, giving the employee a choice of health-benefit plans rather than a single plan.

In order to bring the rather theoretical notion of corporate health-care strategy down to earth, we present four case examples of employers that have been pathfinders in trying to pull the threads together. The cases clarify as well as illustrate some directions worth following. Each case sketches the process of formulating strategy, the design of the strategy, the organizational aspects, and the early stages of implementation.

PFIZER INC.[3]

Formulating strategy

In 1983 Pfizer Inc.'s CEO designated a task group to develop a two-pronged approach to the management of health-care costs for the company's employees in the United States. The task group, comprised of representatives from Corporate Benefits, Employee Relations, and Public Affairs, was charged with managing health costs, promoting employee health, and fostering good employee relations, all at the same time. The two prongs were (1) benefits planning and administration, and (2) health-care cost management at the work site.

With corporate support, the task group made an early decision to try to reduce costs, not by asking employees to share them, but by reducing the *need* for health-care services through the encouragement of behavior that promotes good health.

To begin with, the task group examined other major companies' programs, which, they learned, fall into four general categories: injury and illness prevention, disability management, health promotion, and employee and family assistance.
 —Illness and injury programs are directed toward both work and nonwork environments.
 —Disability management aims to speed an injured employee's return to productive employment through physical and vocational rehabilitation.
 —Health-promotion programs encourage employees and family members to live healthy lives.
 —Assistance programs seek to prevent illness by alleviating the stress that results from personal problems.

The task group, using cost-benefit analysis, estimated that the company's existing safety and health program was saving 25 percent to 30 percent on the total bill for workers' compensation. Also, it found that Pfizer work sites with adequate numbers of safety and health staff had only one-third as many accidents per capita as other sites. Analysis of the other three health-program components in major companies showed cost-benefit ratios of 1-11 for disability management, 1-2 to 1-5 for health promotion, and 1-7 to 1-17 for employee and family assistance.

Design

Pfizer decided to tailor a health-care management strategy to its own needs, with elements from all four categories of programs. The company laid out 10 objectives to be attained through its strategy: improved employee productivity and performance, improved employee health, reduced accidents and illnesses, reduced health-care costs, improved employee relations, improved facility-community relations, a strengthened facility affirmative-action program for the handicapped, reinforcement of the company image as a leader in health care, improved receptivity to benefit changes, and reduced legal liability.

Illness and Injury. The company already conducted programs of hazard control, fire prevention, accident investigation, work-site inspection, right-to-know measures, sanitation control, human-factors engineering, and preemergency planning. To these would be added: family and home safety, motor vehicle safety, sports and recreation safety, and consumer products alert. The decision reflected the discovery that accidents at home cost the company almost three times as much as accidents at work. To avoid intrusion on privacy, however, the programs would concentrate on preventive education.

Disability Management. Since an employee's disability costs the company health-care dollars, lost productivity, and disruption in the work site, regardless of where the disabling accident occurred, the company decided that it would actively manage all cases of disability, work-related or not. Active management includes early intervention, rehabilitation, and selective placement. Intervention begins within a day or two after the accident, with the goal of returning the employee to work as soon as possible. Rehabilitation includes accommodating a job physically to the disabled person's needs. Selective placement helps the patient's physician understand the work site and the content of the job, and thus diagnose the employee's ability to return to work.

Health Promotion. Since 65 percent of health-care costs are incurred because people don't take proper care of themselves, the company decided to provide screening, health-risk assessment, and self-care training.

Employee and Family Assistance. The company contracted with a local service to counsel and support troubled employees and family members on a 24-hour basis through a telephone "hotline." Utilization is voluntary, confidential, and cost-free to the employee. Services include (1) short-term counseling and referral for a wide range of problems, and (2) follow-up counseling and treatment.

Organization

In order to carry out the strategy as flexibly as possible in this multiplant company, management allocated responsibilities between corporate and work sites as follows:

Corporate staff would (1) develop program guidelines, (2) provide information and technical resources, (3) assist with training, needs-analysis, program design and execution, identification of external resources, evaluation, and monitoring of progress.

Work-site managers would (1) develop and administer programs, (2) train their own staff, (3) engage external providers of services, (4) publicize their programs, (5) monitor program usage, and (6) evaluate program effectiveness.

Upon launching the strategy, the corporate task group discovered that it had overlooked an essential element: health-cost accounting. Under existing arrangements the cost data received by the work site were useless for health-care management; so useless, indeed, that work-site managers were unaware that there *was* a health-care cost problem. Health costs were merged with other employee benefits in the fringe-benefit account and were treated as uncontrollable corporate overhead. Each facility was assessed at average corporate rates regardless of the actual costs it incurred. Under these conditions, how could corporate management persuade the facilities that they should allocate time and money toward managing health-care expenses? As of early 1985, therefore, the corporation began communicating actual local health expenses ("expenses" because Pfizer is now self-insured) by memorandum to local managers. In 1985 these expenses began to appear as separate line items on facility expense reports.

For purposes of self-control, the corporation has developed techniques for evaluating its health programs, using four kinds of statistics: financial data, injuries and illnesses, program usage, and population health. A yearly survey will assess what effect the strategy is having on Pfizer employees' health.

CIBA-GEIGY[4]

Process

In April 1981 the president of CIBA-GEIGY appointed a steering committee and a task force to recommend a strategic health-care plan for the company, designed to improve the cost-effectiveness of health-care benefits and the health of employees and dependents. The more detailed goals of the strategy were to contain future increases in direct health expense; maintain and improve the competitiveness of the health-care plan; make quality medical care accessible to employees and dependents; monitor the delivery of health services; and promote healthier life-styles among employees and their dependents.

Certain axioms were adopted. (1) Current levels of coverage would not be reduced, and eventually there should be a consistent level of coverage for all salaried employees; (2) health benefits should attract and retain employees; (3) there should be an aggressive long-term strategic approach to health-care cost effectiveness; (4) management should be prepared to invest money now toward long-term reduction of costs.

Implementation of most of the plan began in January 1983. The remainder was implemented in April and July 1983. A measurement and evaluation system is in preparation.

Design

The plan has six main components:
- ☐ Medical-benefit plan design and administration includes eligibility determination, coordination of benefits, and alternatives to inpatient services (e.g., ambulatory surgery, preadmission testing).
- ☐ Health services/health improvement includes disease prevention, risk assessment, employee education, and health promotion.
- ☐ External involvement includes coalitions, hospital board memberships, and legislative advocacy.
- ☐ Cost sharing includes a slight increase in premium sharing and incentives for preventive dental care.
- ☐ Communications include an explanation of changes in the plan, of reimbursements, and of how the employee fits into the plan.
- ☐ Measurement/evaluation

Further details of the strategy follow.

☐ The medical-benefit plan includes expanded access to alternative delivery systems; encouragement of ambulatory surgery, where appropriate; prevention of duplicate reimbursement for expenses covered by two or more plans; improved determination of eligibility for benefits; retrospective and concurrent utilization review; more preadmission testing before elective hospitalization; predischarge planning, with attention to hospice care; taking advantage of discounts for prompt payment of providers; mandatory second opinions for the most abused procedures; and improved subrogation/reimbursement recoveries.

☐ Health services/health promotion includes an array of communications devices to raise awareness of life-style and health; physical exams to identify health risks; encouragement of preventive measures; support for physical fitness and exercise; encouragement of employees to consult site physicians, whether or not the medical problem is work-related; disability review to ensure a speedy return to work; an employee-assistance program at each site.

☐ External affairs includes participation in legislative debates and information sharing; support for coalitions and similar groups; and education of hospital trustees.

Organization

Primary responsibility for executing the strategy lies with the vice president of human resources, assisted by a task force. Additional resource people will be made available: a temporary part-time medical consultant to guide relations with the medical community; a coordinator of implementation for the strategy; and a health educator. Claims staffing and processing may be reorganized.

The task force has developed several indices with which to measure outcomes of the strategy: overall health cost per employee; administrative savings versus targets (the difference between charges submitted and benefits paid); improvements in utilization (e.g., hospital admissions per 1,000, average length of stay); reduction of health-harming habits and of health risks; reduction in disability days at each site; comparison of CIBA-GEIGY's cost increases with those of other companies; number and type of personal health-care services provided by site physicians.

OWENS-ILLINOIS[5]

Owens-Illinois in 1981 adopted a health-care cost-containment goal: To slow the rate of health-care cost escalation without compromising quality of care and to protect the health of its employees adequately, equitably, and on a sound financial basis. To reach the goal, Owens-Illinois developed a nine-point strategy:
—Adequate funding to implement cost containment
—Designing appropriate incentives into the health-benefits plan
—Educating employees to use benefits cost-effectively and to strive for physical fitness
—Utilization review
—Supporting participation in alternative delivery systems
—Encouraging employees to participate in the management and direction of hospitals, HSAs, and similar organizations
—Supporting community-wide health-planning organizations
—Participating in national and state legislative surveillance
—Participating in coalitions

In 1981 the company's policy committee adopted the strategy and agreed to provide financial and personal support. In June 1982 a Health Care Policy Department was established, to carry out the strategy.

By way of implementation, the following steps have been taken:

☐ The health-benefits plans were redesigned to help employees become more prudent purchasers of services. For example, deductibles and monthly payroll contributions were raised; incentives were modified to encourage outpatient testing and surgery, at-home care, and treatment at Emergicenters; and a second-opinion program, administered by registered nurses in a third-party health agency, was made mandatory for elective surgery.

☐ Wellness and prevention programs were introduced, with many easily accessible classes, workshops, and seminars, and free and frequent work-site screening programs. The registered nurses mentioned above provide advice on outpatient surgery, preadmission testing, and rehabilitation programs.

☐ The company gives financial support to HSAs and other agencies concerned with controlling capital costs in the health-care delivery system.

☐ The company is an active member of seven coalitions,

located in communities where large groups of its employees live.
- ☐ The company has sponsored seminars to educate hospital trustees and has encouraged its employees to serve as trustees.
- ☐ The company plans offer membership in HMOs, and over 20 percent of its employees are enrolled.
- ☐ Utilization review is conducted by PSROs under contract and also by the company's medical staff.
- ☐ The company has been active in groups organized to influence national legislation on health care.

By 1983, 25 percent of the work force was covered by plan changes; the other 75 percent were covered in 1984. Cost containment has improved sharply, primarily as a result of the mandatory second-opinion program and case management by registered nurses. Employee surveys have indicated wide approval of these programs, and their scope has been extended.

GENERAL ELECTRIC[6]

In 1982 GE created a Health Care Management Team to pull together all the health-care activities of this highly decentralized company. The team was bicameral, including both the Corporate Council, with representation from many corporate staff departments, and the Health Care Network, with 148 health-care coordinators from the work sites, plus key area coordinators. The goal was to manage health-care costs while maintaining or enhancing benefits.

Two company-wide conferences were held, one in 1983 and one in 1984, to enable work-site managers to exchange experiences and ideas at first hand. Although work sites are encouraged to develop their own approaches, the corporation promotes health care along seven lines:

—GE's data base, containing data on employee and provider costs and services, as well as diagnosis and procedure coding
—Claims review
—Gradual introduction of utilization review at work sites
—Consumer education and communication
—Continued support of The Business Roundtable Health Initiatives project

—Review of the programs of all of the company's 250 plants, and visits to selected sites
—Corporate-wide evaluation of local programs

By 1984 most divisions had submitted their first plans for corporate review, and several field reviews had occurred.

GE's Pittsfield, Massachusetts, plant exemplifies local initiatives. The local managers were mainly concerned about hospital admissions and stays, utilization rates for certain surgical procedures, short-stay admissions, utilization of alcohol abuse programs, and weekend hospital admissions. In response, they adopted preadmission screening, concurrent review, discharge planning, and second opinions for elective surgery, with objectives set for each program. A weekly sickness and accident tracking system was adopted, and an EAP was begun in 1984. A medical utilization review committee, consisting of nine local physicians and four GE managers, meets monthly; they examine absenteeism, peer review, and hospital stays. A health fair conducted by the plant drew 20,000 people, and many important screening tests were administered. Union cooperation in the program has been strong.

The foregoing cases should suggest guidelines for employers who feel the need of a strategy to unify their health-care management. Those wishing further details are advised to contact the companies themselves. The cases make clear that a successful strategic approach requires thoroughness in design and execution, and willingness to invest time and money.

UNION STRATEGIES

Unions, no less than corporations, have found it desirable to develop unified approaches to the complex problems of health-care management. Working within tighter constraints than employers, as a rule, they have come up with diverse and sometimes ingenious solutions. United Auto Workers for example, deals with some of the largest companies in the world, each of which has its own health-care arrangements; accordingly, the union strategy features policy analysis, advocacy, and the design and establishment of alternative delivery systems. International Ladies' Garment Workers' Union and the Teamsters, on the other hand, deal with a multiplicity of not-very-large employers and therefore operate direct services for members and their dependents. All three of these unions faced up to health-care cost containment long before their employers did.

United Auto Workers (UAW)[7]

UAW's leaders were aware even in the 1950s that cost controls were vital to their hopes of expanding health-care benefits. By 1964 the union was urging Michigan's insurance commissioner to integrate hospital service planning, budget review, and payments to improve long-term care facilities, to expand alternative delivery systems, and to establish peer review committees. It urged a study of hospital charges and opposed Blue Cross's effort to replace community rates with experience rating.

In 1968, in testimony before the U.S. Senate Committee on Government Operations, Walter Reuther pointed out that the cost-based system of reimbursement created no incentive for cost efficiency and that the nation's aim should be to use quality standards and incentives to keep people well and out of the hospital. He advocated financial aid to prepaid group practices, the regulation of hospitals as public utilities, and a host of other regulatory measures to improve the health-care system.

The union paved the way for the HMO movement in the 1960s by developing comprehensive specifications for prepaid group practice: monthly premium payments in advance to cover preventive and remedial services; services to be provided by full-time physicians whose incomes are unrelated to the number of patients treated or fees received; selection of staff physicians by their peers; group accountability and review of procedures; common laboratory and other services; availability of social and psychological counseling; and modern technology. From the beginning it opposed cost shifting, on the ground that the consumer had no control over health costs.

More recently, UAW has attacked the big differences in cost of treatment in different hospitals and the unexplained local variations in the use of surgical procedures and in-hospital stays for the same conditions.

Other pace-setting cost-control programs adopted by UAW include:
—Preventive dentistry
—Coordinated home care as an alternative to hospitalization
—Hospices
—Case management by peer review

—Hospital utilization review, mandatory second opinions, preadmission testing
—Home care for hemophilia
—Predetermination for podiatry and dentistry treatment
—Ambulatory surgery, with associated changes in sickness and accident policies.

At the same time that UAW works actively at creating HMOs and PPOs, it screens them vigorously for compliance with quality standards.

International Ladies' Garment Workers' Union (ILGWU)[8]

ILGWU offers its members two alternative forms of health-care benefits, both financed by collectively bargained contributions from the small employers who make up this beleaguered, low-wage industry. On the one hand, through the union, all members have Blue Cross/Blue Shield coverage equivalent to that of any other card holder. On the other hand, the union maintains a system of four big-city union health centers.

The center in New York City is the largest free-standing ambulatory service in the United States. Beginning in 1911 as a two-room office where a physician examined applicants and certified members for sick benefits, it grew into a diagnostic clinic in 1914. In 1919 it moved into larger quarters and became a cooperative medical center, financed by local dues. Since 1945, when it moved to its present location on Seventh Avenue near the garment district and expanded again, the center has been financed by multiemployer payments.

A large, modern group-practice facility, the center serves members, dependents, and retirees from the city and suburbs. Services are free except for a nominal fee per visit (even for a CAT scan).

The range of care approaches that of an HMO, with a staff of 85 physicians in 30 medical specialties, as well as nurses, technicians, patient helpers, social workers, clerical workers, and a pharmacy. (The union's mail-service program fills 2 million prescriptions a year at discount rates.)

In addition, the center provides such preventive and wellness services as:
—a three-hour complete medical examination
—active promotion of exercise, nutrition, smoking cessation, regular check-ups, mental-health care, stress management, and preventive screenings

—monthly union newspaper articles on staying healthy
—an employee-assistance program (personal and family, occupational alcoholism, and health counseling).

Teamsters[9]

Teamsters Joint Council No. 16 Hospitalization Trust Fund has coordinated the medical and hospital needs of Teamsters welfare funds in New York City since the late 1950s. Guided by the Columbia University School of Public Health and Administrative Medicine, the fund in 1958 carried out an extensive study of health-care costs and quality; about 100 hospitals allowed it access to their records. The survey found that one-fifth of hospitalized members received poor care and one-fifth of admissions were unnecessary.

After lengthy planning, the fund joined with Montefiore Hospital and the Columbia University School of Public Health and Administration to form the Teamster Center Program. Four units were created to serve members and dependents:

- ☐ The Diagnostic Unit, for second opinions in cases requiring complex and serious surgery.
- ☐ The Medical Advice Unit, to help employees choose the best sources of care—selecting a physician, referrals to community resources, and resolving medical and financial problems.
- ☐ The Treatment Unit, to treat members suffering from the most serious diseases.
- ☐ The Medical Audit Unit, to review and evaluate the quality of hospital and medical care received by members.

A second Teamster Center was established at Long Island Jewish Medical Center in 1969. Both continued until 1978, when the widespread adoption of major medical coverage by the participating welfare funds made the centers redundant. They were replaced in the same year by a more limited and less costly agency, called Teamster Center Services (TCS).

Essentially an advice and referral unit, TCS serves 32,000 members and their families in New York City and the surrounding area. Members receive guidance on alcoholism, drug abuse, vocational reeducation, mental health, medical insurance, legal and financial matters, child and senior care, and second surgical opinions (19 percent of which have indicated no surgery needed).

TCS visits community resources periodically, drops poor performers from the referral list, and adds good ones. It has overcome members' resistance to second opinions on surgery by means of educational programs, a hot line, and a list of hand-picked consultants. A newsletter advertises TCS's services and provides health-related advice.

Most of TCS's services are performed over the telephone, a method that is user-friendly and cost-effective. If personal visits are required, an internist is available at Montefiore Hospital by appointment. In addition, TCS runs health-risk-reduction workshops, a hypertension control and health promotion program, and an employee alcoholism assistance program.

RECOMMENDATIONS

20 Employers and unions intent on improving health-care management should develop a strategic framework to coordinate their action programs. A corporate strategy makes it possible to avoid omissions and overlapping and ensures that the various programs tend in the same direction.

21 A corporate health-care strategy should include the following elements:
—A review of the organization's experience in health care, forecast of its present tendencies, and selection of targets for improvement
—A survey of the organization's current data sources and an estimate of those it will need
—Practical goals for improvement
—Assessment and redesign of the organization's health benefits
—A plan for managing health benefits more effectively
—For each aspect of the strategy, an assignment of responsibilities and an allocation of staff and financial and data resources
—Active, visible support from top management

NOTES

1. Hewitt Associates, "Employers Tighten the Administrative Reins on Medical Plans, *News and Information*, October 29, 1984.

Victoria George and William E. Hembree, *Breakthroughs in Health-Care Management: Employer and Union Initiatives*. Pergamon Press/Work in America Institute Series (New York: Pergamon Press, 1986).
2. The Business Roundtable, *An Appropriate Role for Corporations in Health-Care Cost Management* (New York: The Business Roundtable, 1982).
3. George and Hembree, *Breakthroughs in Health-Care Management.*
4. Ibid.
5. Ibid.
6. Ibid.
7. Ibid.
8. Ibid.
9. Ibid.

8. REDUCING EXCESS HOSPITAL CAPACITY

Much of this report is devoted to the goal of reducing avoidable hospital utilization, because hospitalization is the largest single component of health-care costs in the United States. Roughly 40 percent of the nation's health-care dollars go to hospitals, and more than 60 percent of the average company's health-care dollars are spent in hospitals.[1] However, efforts to reduce hospitalization will come to very little unless the grossly excessive numbers of hospital beds in the United States are lowered at the same time in an orderly and considerate manner.

Excess Hospital Capacity as a Cause of Higher Costs

A major reason why hospitalization costs are so high is, paradoxically, the fact that hospital capacity (measured by the number of patient beds and employees per capita) far exceeds what is needed. The amount spent on hospitalization is a function of the number of admissions, the length of stay, and the cost per day of hospitalization. Excess capacity has caused all three of these factors to be higher than they would otherwise be.

In a free market, excessive capacity leads to distress selling, closures, and an eventual equilibrium at a lower level of capacity. In health care, however, inflated demand, which has prevailed for many years, masked the excess until the mid-1960s. Hospitals stayed afloat by stimulating physicians to increase the number of admissions and length of stay and by raising per-diem charges to offset fixed overheads. Some hospitals are able to break even with as little as 45 percent occupancy.

The strength of the relationship between excess capacity and health-care costs is shown by Professor Wennberg's discovery that although health-care costs and frequency of hospital admissions per capita vary widely from one area to another, the most reliable predictors of higher frequencies of admission and higher costs per capita are (1) number of hospital beds per capita, (2) number of physicians per capita, and (3) number of hospital staff per capita.

How Excessive Is Capacity?

Nationwide hospital occupancy dropped in 1982 for the first time in over 30 years as employees and the federal government began to apply pressure. Between 1983 and 1984 nationwide admissions went down by 3.7 percent; the length of the average patient stay, about 5 percent. In Iowa, Blue Cross/Blue Shield subscribers reduced their use of hospitals by 16 percent over three years, leaving many hospitals half empty and financially weakened. These declines—due not only to changes in reimbursement and the development of alternate delivery mechanisms, but also to the effects of the 1982 recession and massive unemployment, which left many individuals without health coverage—barely hint at the true excess.[2]

A large body of experience leads consistently to the conclusion that, for the United States as a whole, current hospital capacity exceeds real requirements by 20 percent or more. "Exceeds real requirements" means that the nationwide hospital capacity of staffed and open beds could be reduced by 20 percent without impairing the quality of health care. It is not within our scope to guess how far such a reduction would lower the nation's expenditures on health care, but it could not help but lower them significantly and permanently.

Let us review the evidence:

1. An authoritative study by Harold Luft in 1981 showed that enrollees of prepaid group HMOs have 10 percent to 40 percent lower total health costs than enrollees of fee-for-service plans. "Lower costs are due mainly to lower hospital utilization . . . There appears to be a 30 percent lower use of hospital days for PGP (prepaid group practice) members and 20 percent fewer for IPA (independent practice association) members. Hospital admissions for PGP subscribers are 20 percent to 40 percent lower than for people in conventional plans" (see chapter 6, "Health Maintenance Organizations").

2. In 1982, enrollees of all ages in all types of HMOs used 458 days per thousand, which is 60 percent less than the 1,134 days per thousand for all short-stay hospitals in 1981. Although HMO enrollees over the age of 65 used 2,084 days of hospitalization per thousand in 1982, this is still 52 percent less than the 4,382 days per thousand for non-HMO people over age 65 in that year.[3]

3. The Rand study, covering the period 1974-1982, found that hospital admissions for HMO enrollees without cost sharing were 40 percent less than for people in fee-for-service plans without cost sharing. "The reduced use of hospital services was found in both the group randomized into the HMO and the group already enrolled there, implying no selection of risks for or against the HMO in this case."[4] All of the above studies found that the quality of health care received by HMO enrollees was identical to that received by members of conventional plans.

4. The evidence also shows that even under a fee-for-service plan, hospitalization can be reduced without impairing quality of care. For example, another finding of the Rand study was that hospital admissions among adults in fee-for-service plans were some 30 percent to 50 percent higher when all care was free than when there was cost sharing. (Admission rates for children were unaffected by cost sharing.) "Among the fee-for-service plans, the number of preventive visits was higher in the plan with free care than in the plans that required cost sharing, but the hospitalization rate was higher, not lower, in the free care plan."[5] The clear implication is that the more frequently an individual makes preventive visits to a fee-for-service physician, the more likely the physician is to see illness that leads to hospitalization, with no gain in the individual's health to show for it.

5. Employers whose plans include mandatory second opinions or preadmission certification report reductions of 10 percent to 20 percent in the rate of hospital admissions.[6] Significant reductions also occur in plans that provide incentives for ambulatory, as opposed to inpatient, examinations and treatment.

6. The Wennberg small-area statistical studies demonstrate wide variations in the frequency with which physicians in different areas admit patients to the hospital for a given diagnosis. When physicians in high-frequency areas are made cognizant of the findings, their style of practice converges fairly rapidly toward the average.[7]

A nationwide excess capacity of 20 percent does not necessarily mean that every community has 20 percent too many hospital beds. Some may have more; some may have fewer. However, in the opinion of a leading health planner, it is "logical to assume that almost every community is 'overbedded.' "[8]

The excess came about as a product of several interacting forces: rapid changes in technology that allow many more procedures to be successfully performed out of the hospital; restrictions in health insurance and perverse incentives built into deductibles and coinsurance; the economic convenience of the hospital as a setting for physicians; an oversupply of government and charitable funds for hospital construction; competitive pressures among hospitals and physicians; insufficient participation by employers and unions in the work of official health-planning agencies; and misguided participation by employers and unions as trustees of hospitals. None of these forces justifies the continuation of excess capacity, now that the impact on health-care costs is understood.

Despite the excess, pressures to *expand* capacity persist in many places—even in the State of Michigan, which is financially strapped and overbedded and has a declining population. Efforts by a statewide employer-union committee, the Economic Alliance of Michigan, to stem the tide have been frustrated thus far in the courts and legislature by physician and hospital representatives. Almost everywhere, legislative and regulatory devices, such as the certificate-of-need procedure, have failed to contain the expansion. According to a national study by Arthur D. Little, Inc., every dollar invested in hospital capital between 1974 and 1978 raised annual operating costs—over and above the costs of amortization and interest—by 16 cents for nonprofits and by 30 cents for others. Business and union leaders who serve on the boards of hospitals and foundations should stand firm against any expansion that cannot be objectively justified.

Where regulatory measures have failed, raw economic forces have begun to make inroads. Employers' and unions' new stress on cost containment, the federal and state adoption of prospective payment, and poor economic conditions have taken their toll in several communities. Hospitals have been forced to reduce capacity, limp along, or go out of business altogether. In such cases reduction has taken place with little or no attention to the welfare of the community or of hospital employees. All these strands came together in the widely publicized Minneapolis nurses' strike in 1984.[9]

The Hazards of Unplanned Reduction

When excess capacity is reduced *ad hoc* in response to economic pressures alone, the potential gains to the health-care system and to the economy in general may be dissipated. The hospitals that close down are not necessarily those which the community, if it thought about the situation objectively, would want closed. Strong political currents are stirred among the multiple stakeholders of the affected hospitals. Unplanned closures also raise havoc among hospital workers, imposing personal hardships and sharpening political resistance.

Economic Pressure vs. Community Needs. The growth of alternative delivery systems and the changes in reimbursement systems will increasingly threaten the continuation of excess capacity in communities. Since the reimbursement mechanisms will no longer "prop up" unneeded capacity, financial difficulties and even bankruptcy will face those hospitals unable to control costs and/or maintain high levels of occupancy. The local hospital's insolvency will, in turn, create problems for the community.

There is no assurance that hospitals which are unable to cope are the right ones to close or convert to another use. For example, often the hospitals with the fewest options in the tightening financial environment are public hospitals—inner-city nonprofits serving large, poor, and minority populations, or rural hospitals. Each type meets a special need, and their demise makes the delivery of health services to their user groups less convenient and more costly. The focus on financial "winners and losers" in the debates concerning reimbursement policies has obscured the more important creation of community "winners and losers" that result from those policies.

Interest Groups. The interests of the hospital and its stakeholders are both complex. Any move which reduces a hospital's size, service, mix, prestige, or its attractiveness to physicians, to name a few factors, will affect one or more of the stakeholder groups. To gain a clear grasp of the potential reactions to capacity reduction, the stakeholders and their interests may be identified as follows:

Internal Stakeholders

Trustees or other governors
Administration
Medical staff
Employees and unions

Auxiliaries
Health-professions training programs
External Stakeholders
Patient population
Third-party payers
Regulatory agencies
Planning bodies
Community organizations and consumer groups
Suppliers
Health-professions training programs
Neighborhood associations
Ethnic societies
Religious groups
Employers
Unions

Other groups and subdivisions could be added to each list, but the examples suffice to show the complexity of the typical stakeholder mix.

To judge from case histories of mergers and closures, any decrease in capacity for services or in number of beds, even short of closure, generates powerful stakeholder reactions, ranging from legally based claims of discharged employees to emotion-laden protests of communities losing "their" hospital. The intensity of reaction reflects the complexity of the problems involved— the flight of wealthier hospitals to the suburbs; the loss of admitting privileges for physicians; increased cost or difficulty in reaching a hospital; the need for standby capacity to cope with emergencies; the need for emergency services even when they are not fully used.

Hospital operations are labor-intensive. In smaller communities, the hospital may be the largest single employer. In larger communities, the hospital employs many people in low-skill positions, such as housekeeping, maintenance, and food service, who would have difficulty in securing similar employment outside the hospital field. Major reductions and closures hit them hard. The employees' reactions and the hospitals' handling of employee-related aspects of ad hoc capacity reduction have been studied by the Orkand Corporation and by Lewin and Associates. The Orkand study was case-based; the Lewin study was based on a literature review.

Orkand's study of eight closed or closing hospitals examined such employee-related concerns as: advance notice, payment of wages due, payment of accrued vacation time, payment of unused sick time, pensions, job-placement assistance, severance pay, and continuation of health benefits. After extensive interviews with former employees of the eight institutions, the investigators concluded that:

- ☐ Ad hoc closure of a hospital is a chaotic experience.
- ☐ Special financial assistance to hospitals that are closing does not, of itself, ensure reemployment for displaced employees.
- ☐ Community organizations, such as health systems agencies (HSAs) and local hospital associations, generally play a minor role in assisting displaced employees.
- ☐ Salaries of new jobs for displaced employees are generally lower.
- ☐ The ability of displaced persons to find new jobs varies.
- ☐ Employees of the closed hospitals felt that they were not fairly dealt with.

Several of the Orkand findings, however, indicate that such problems can be ameliorated by care and planning:

- ☐ If the hospital closes *before* total financial collapse, the closure is more orderly and employees are treated more equitably.
- ☐ Hospitals that receive financial assistance during a closing are able to close down in a more orderly fashion, providing advance notice and help with employee placement.
- ☐ Union leadership can provide job-placement assistance, and can keep the hospital management's attention focused on employee issues during the closure process.
- ☐ Job-placement activities by outside entities are generally more effective than those of the closing hospital.
- ☐ Subsidies encourage "receiving" hospitals to hire displaced employees.[10]

The Lewin study, in which employee concerns figure only as one societal factor relating to hospital capacity reduction, points out that the hospital industry employs proportionately more blacks, Hispanics, and women than other industries. This is attributed to traditional low salaries in the field, which have tended to attract those unable to find other employment; the nature of many of the "hotel" service jobs in the hospital, which require neither training nor great strength; and the rapid growth of employment in the industry at the same time that large numbers of minorities have entered the work force. Thus the potential unemployment associated with closures disproportionately affects a vulnerable group.

A second employee-related factor identified by the Lewin study is the degree of unionization in a closing hospital. Unions have opposed closings effectively, but they have also facilitated them through job-placement activities, depending on the specific situation.[11]

A Role for Employers and Unions

Excess capacity is usually approached as a "system" problem, national or statewide in scope. Health-care professionals use the metaphor of squeezing the balloon at one point, only to see it expand by the same amount elsewhere. But that does not mean that the problem of excess capacity can be solved by attacking "the system." There is no system, in that sense. Federal and state governments can shrink the system by making less money available to it, but they cannot control the when, where, and how of local shrinkage.

If the objective is to effect an orderly reduction of capacity, it is more realistic to think of the national system as a collection of local systems than to regard the local systems as subdivisions of the national system. Hospital costs in each community reflect the local demand and supply of hospitals, and, as Wennberg and his colleagues have demonstrated, such costs vary widely from one community to the next.

In political terms, employers and unions can exert far more clout locally than at state or federal level. State and federal planning of hospital capacity has been ineffectual because planning agencies have had no constituencies to back them, whereas providers have wielded power with skill and determination. At the community level, however, the tables are turned. Coalitions of employers and unions can direct (and have directed) their purchases of health care in such a way as to compel reductions of capacity. Employers and unions are in a strong position to exert leadership through actions related to their direct interests and through philosophical and political support for broader steps.

Direct action might include the restructuring of health-insurance benefits to encourage local providers to deliver care more effectively to plan beneficiaries; service by business and labor representatives as hospital trustees to assure that management and medical staff members are actively pursuing more efficiency in the delivery of services, and to support capacity reduction where indicated by community and institutional factors; assistance in job placement for displaced hospital workers; and financial and personnel support to organized community efforts to reduce capacity.

But in the absence of a well-constructed community plan, much of their effort may be wasted. In the early days of hospital capacity planning, when the aim was expansion, employers and unions often joined forces with planning agencies. This pattern fell into disuse during the period of most rapid expansion, when the readiness of federal money made growth seem easy and cost-free. Now that the interests of planners, employers, and unions again coincide, the alliance needs to be revived.

Where there is no community-wide, hospital-specific strategy for bed reduction—that is, in most areas—an appropriate role for business and labor is to facilitate and support the development of such a plan. (In 1985, for example, the Milwaukee Association of Commerce engaged a well-known consulting firm to draft one.) Often they can succeed where health planners have lacked the power to do so.

As parties with a major interest in the community's health system, both from the standpoint of its costs to them and its value in maintaining the quality of life in the community, employers and unions are beginning, in some communities, to take the leadership in planning for reduction of excess capacity and in implementing those plans.

Contents of a Plan

To ensure practicability, a community's bed-reduction plans must take account of questions such as: Which hospital's capacity is to be reduced? By what means? What help will be required by those that reduce? What help will be given to displaced employees of those hospitals? What are the options when a changing environment creates excess hospital capacity? How can the "sunk capital" of redundant physical plants be protected and made productive? The options are relatively simple:
—Closure of entire institutions
—Closure of nursing units in institutions
—Consolidation or merger, with reduction in bed capacity
—Conversion to another use
Closure may be voluntary or involuntary. In either case, the

facility ceases to perform the functions of a hospital, its employees lose their jobs, its physicians lose their workplace, and the community loses a source of care. From an economic perspective, closure of entire institutions and removal of their beds from the community's supply are the most effective means of reducing the overhead costs associated with that capacity. From a social and political perspective, this approach is the most traumatic, however.

Partial closure, where the institution discontinues a specific service involving beds, is frequently touted by hospitals as a solution to an actual or perceived excess. From the hospital's perspective, partial closure is better than total closure for a number of reasons: the organization is preserved and continues to function, hope of eventually reopening the unit or units is not lost, and the burden is spread. It does not, however, offer the cost-saving advantages to the system that a complete closure does.

Consolidation or merger (that is, an institution ceases to exist as an independent entity, as its assets are transferred to another entity) is currently the most popular option. But, bed reduction is often not the primary motivation. Improvement of the involved institution's financial position often outweighs bed reduction as an incentive. In this context, merger is defined as the situation in which the key advantages sought are those of efficiency, service, quality, and process. Mergers can be a useful tool in reducing capacity if the surviving institution has the resources to assist the closing institution and its employees, and if the result is a reduction in capacity.

Finally, beds may be removed from the system when a hospital is converted to another use. If the hospital plant becomes an apartment complex or an office building, it is essentially a closure, but if the hospital management redirects the use of its facilities to another health purpose, such as long-term care, capacity has really not been reduced.

The sale of existing institutions to other community facilities may have detrimental effects from an antitrust perspective and may, unless monitored and guided by interested outside parties, result in capacity reductions in those portions of the community which need the services the most.

Assistance to Hospitals That Reduce Capacity

The Orkand study identified three elements needed in a program to reduce capacity through closures: flexible financial assistance to the closing hospital; comprehensive job-placement assistance from an outside entity and incentives to neighboring hospitals to assist through hiring; and informal merger arrangements to ease the impact of closure. Employers and unions could cooperate in creating the necessary structures and in organizing the assistance networks.

An organized effort outside the hospitals themselves to find employment for displaced workers is important, not only for the workers, but also as an aspect of managing the impact of closure on the community. Employers and labor should ensure that the interests of the community at large are respected in the process of capacity reduction, especially those groups least able to represent their own interests.

The Process of Planning

The effects of closure on the community are largely dependent on the condition of the closing facility and the way it manages the closure process. If the need for capacity reduction can be brought into the open, discussed with all interested parties, and incorporated into an institution-specific plan, the prospects for community acceptance and adjustment are enhanced. Employers and unions can provide the forum for such efforts in conjunction with planning agencies and other community interest groups. Health-systems agencies, where they exist, should be able to provide help in determining the starting points.

RECOMMENDATIONS

22 Employers and unions should apply their clout at community level to reduce hospital capacity to the optimum, since the existence of excess capacity counteracts all efforts to lower the cost of hospitalization. The optimum, at present, is the capacity that would be required if prepaid group HMO hospitalization rates were the norm for the community.

23 Employers and unions, in collaboration with other civic leaders and interested parties, should ensure that hospital-capacity reduction proceeds in accordance with an openly-arrived-at, community-wide, hospital-specific plan. If such a plan does not exist, employers and unions should work with local hospital-systems agencies to ensure that one is developed and carried out.

24 Employers and unions should ensure that the community plan provides for the needs of the indigent and the underserved. Hospitals that serve these sectors of the community tend to be the weakest financially, but their closure may cause widespread distress and a political furor that undermines further efforts to reduce capacity.

25 Employers and unions should ensure that the community plan provides for hospital employees to receive suitable advance notice of closures and organized assistance in obtaining new jobs. Nonmanagement employees of hospitals suffer the most in a closure, although they are the least to blame. They have the smallest resources to fall back on and the greatest difficulty in finding replacement jobs.

NOTES

1. "The Corporate Rx for Medical Costs," *Business Week*, October 15, 1984, p. 139.
2. "Iowa Health Program Trims Insurance Fees," *The New York Times*, June 14, 1984, p. 1.
 "Decline in Hospital Use Tied to New U. S. Policies," *The New York Times*, March 16, 1985, p. 1.
3. James R. Kimmey, "Dealing with Excess Hospital Capacity: Methods and Consequences," an unpublished paper commissioned by Work in America Institute, October 16, 1984, p. 10.
4. Joseph Newhouse, "The Health Insurance Study: A Summary," an unpublished paper commissioned by Work in America Institute, October 16, 1984, p. 3.
5. Joseph Newhouse, "Summary of the Rand Health Insurance Experiment," an unpublished paper commissioned by Work in America Institute, October 16, 1984, p. 6.
6. O. David West, "Pre-Admission Certification: A Golden Opportunity to Start Managing Health Care," an unpublished paper commissioned by Work in America Institute, May 9, 1984, p. 4.
7. John Wennberg and Alan Gittelsohn, "Variations in Medical

Care among Small Areas," *Scientific American*, April 1982, pp. 120-134.

Daniel F. Hanley and David N. Soule, "The Maine Medical Assessment Program: Informed Inquiry by Practicing Maine Physicians into Common Medical and Surgical Treatments," an unpublished paper commissioned by Work in America Institute, January 9, 1985, p. 2.

8. Kimmey, "Dealing with Excess Hospital Capacity," p. 12.
9. J.L. Drake, "Toledo Compromise Settles Construction Standoff," *Business and Health*, June 1984.
10. Orkand Corporation. *Hospital Closure: Findings, Conclusions, and Recommendations on Employee Issues and Considerations for Facility Conversion.* (Hyattsville, Md.: U.S. Bureau of Health Facilities, 1980).
11. Lewin and Associates. *Societal Factors and Excess Hospital Beds: An Exploratory Study*, vol. 5 (Hyattsville, Md.: U.S. Health Resources Administration, 1979).

9. CONTROLLING THE USE OF MEDICAL TECHNOLOGY

The rapid increase in health-care costs over the last 30 or so years is in large part attributable to major scientific advances translated into new medical technology and the technology's wide diffusion. While, clearly, these advances have had dramatic positive effects on health status, disability, morbidity, and mortality, there have been negative effects as well.

In many industries new technology—such as computers and electronics—has led to greater efficiency and increased productivity. However, in health care the opposite is often the case. Technology in health care has led to a need for a more and more highly skilled work force (an expansion of physician specialists and numerous technicians). A number of technologies add, rather than substitute for, units of service. In other industries, as technology diffuses, prices tend to fall. Not so in health care. In fact, prices rise.

The development and diffusion of new medical technologies show no sign of slackening, and expenditures on them continue to rise along with popular demand. If allowed to expand unchecked, they will undo much of what employers and unions have accomplished, and what can still be accomplished, to improve the management of health care. Employers and unions must try to get a grip on these developments, using the best advice they can find. Fortunately, there are opportunities for them to do so.

Before the 1930s, the range of medical technology was limited to certain surgical techniques, aspirin, digitalis, and limited laboratory tests. Infectious diseases were the leading cause of mortality, and sulfa drugs and antibiotics were unknown. Major advances in health status were attributable to public health measures of clean water, sewage, vector control, and so on. The average life span at that time was about 50 years.

The research conducted during World War II and the increasing growth of federal funding for biomedical research which continued until the late 1970s spurred a revolution in medical science and practice. The progression included: *drugs*, such as penicillin, the mycins, beta blockers, cyclosporine, chemotherapeutic agents; *devices*, such as lasers, balloon catheters, dialysis machines, heart pumps; *diagnostic procedures*, such as CT scans, nuclear imaging; *treatment procedures*, such as coronary bypass surgery, organ transplants, neonatal intensive care. The computer also has had a major impact on medical technology. Results of these breakthroughs have been that people born in 1983 can expect to live until almost 75. Infant mortality fell from 20 per 1,000 in 1969 to 12.5 in 1981; deaths from cardiovascular illness and stroke have declined dramatically.

Technology Assessment

Despite the undoubted achievements of recent years, there is, according to the U.S. Congress, "an emerging consensus ... that many technologies have been widely adopted into medical practice in the face of disturbingly scanty information about their health benefits, clinical risks, cost effectiveness, and social sideeffects. In addition, the use of some technology persists long after it becomes evident that these technologies are of marginal utility, outmoded, and even harmful.[1]

In 1978, Congress established the National Center for Health Care Technology. The purposes of the center were to set priorities for technology assessment and coordinate activities at the federal level, identify new and emerging technologies, conduct and fund assessments, conduct and fund research into assessment methodologies, and provide advice to Medicare on the safety and efficacy of technology. The life of the center was brief. Funding ended in 1981, and support never reached a level where large-scale assessment activities were possible. The function of providing advice to Medicare was transferred to the National Center for Health Services Research and Technology Assessment (NCHSRTA).

Technology assessment in health care is not a simple process or set of processes for a number of reasons:

□ Technologies change over time; premature assessment can

144 Improving Health-Care Management in the Workplace

result in flawed conclusions, since many technologies evolve through trial and error.
- Diffusion of potentially life-saving technologies can be delayed for a number of years while they are still in investigational stages.
- The safety and efficacy of a technology can vary, once diffused from academic/tertiary settings to community practice settings. Volume and frequency affects efficacy and safety.
- The entry of a new technology does not automatically eliminate existing unproven or less-than-effective diagnostic and treatment modalities.
- The sheer volume of medical procedures precludes assessing all technologies.
- Social and ethical issues related to technologies are extremely complex, and different societal groups will respond differently; abortion and the question of when the use of life-prolonging technology should be discontinued are examples.

The techniques for assessment also have limitations, compounded by the fact that experimentation on humans is often involved. Such techniques also pose problems, such as statistical reliability, the timing of the assessment, control of experiments, and costliness. Thus, other, less rigorous assessment methods, such as synthesis of the literature, often have to be relied upon.

Even when there is sufficient scientific evidence to support the conclusions of assessment, dissemination of the findings to practitioners and the public is slow. For example:

- Assessment of coronary bypass surgery indicates that it is a safe and effective technique for certain types of coronary artery disease. However, its effectiveness for other types of the disease or for some degrees of severity of the disease is questionable, and less expensive medical procedures may be just as effective. Furthermore, morbidity and mortality rates vary widely by hospital and physician. It should be noted that as this procedure has diffused, the actual price has risen rather than fallen, in part because of the way physicians' fees are set.
- Caesarean section rates have been rising. The National Institutes of Health (NIH) held a consensus conference to establish criteria for the use of Caesarean section, but many physicians have not accepted the guidelines.
- Who should pay for the artificial heart or heart transplants?

Fewer than ten implants of artificial hearts in humans have occurred, and none of the patients has survived for as long as a year. In the case of heart transplants, careful assessment shows that the technique is no longer experimental and is effective, with half of the patients surviving five years or more. However, should this technique be available to anyone, or should the Stanford criteria, based on age, medical condition of the patient, and social support systems, be maintained?

☐ How should technology be distributed? Should we continue to encourage decentralization even where use and cost do not warrant the degree of dispersion and where quality questions are raised? How many shock trauma units, transplant units, nuclear magnetic resonance machines, and neonatal intensive-care Level III units should there be and who should decide?

☐ The need for the volume of surgery performed for cataracts, hysterectomies, and prostates has been questioned by numerous medical experts. The United States has among the highest surgical rates in the world, although this has not brought with it the best health-care or mortality rates.

Health-Benefit Plans and Technology

European and other industrialized countries are struggling with similar issues. They have been able to place some controls on diffusion and use, however, through supply control under national health systems. Medical criteria for access to the limited supply of technology are then established regionally or locally, generally by physicians. In Britain, age criteria have also been used for access to such technologies as dialysis for chronic renal failure.

Our decentralized health-care system cannot resolve these issues in the European manner but, faced with limited resources, it must find a way to resolve them. Competition and the market alone are not adequate methods for assessing and diffusing technology. Action will be required not only from federal, state, and local governments but also at the grass roots, from health-benefit plans, their beneficiaries, and providers.

Coalitions can exert significant control on the diffusion of high-cost technologies to hospitals and physicians' offices, using established planning mechanisms and reimbursement policies.

The reimbursement policies and utilization controls of health-benefit plans can be used to shape the diffusion and appropriate use of technologies. The potential of reimbursement has not been

fully realized. Currently reimbursement policies used are limited to the following types of policies:
— Exclusion of experimental technologies from coverage
— Second surgical opinion and concurrent reviews for appropriateness
— Incentives for use of different types of services
— Denial of reimbursement by Medicare, based on recommendations of the National Center for Health Services Research and Technology Assessment that the technology is not safe or effective, or disapproval by the Food and Drug Administration of certain drugs and devices. Some insurance companies also follow the Medicare recommendations.

Reimbursement policies could be tailored to permit more active intervention in technology decisions through techniques that:

☐ Limit payment to facilities designated by the payer as the regional resource for selected high-cost technologies, based on planning-agency decisions and/or quality and volume criteria.

☐ Reduce payments for services as the technology is proven effective and diffuses, as actual costs decline, and as physician supply and skills increase. Revision of numerous surgical procedures, for example, may reduce marginal use of technology and allow for the substitution of less costly approaches. For example, fees paid for Caesarean section are substantially higher than for normal childbirth or safe forceps delivery. The reevaluation of fees would reduce the incentive to perform the procedure in many cases. A similar approach could have spurred the shift from radical breast surgery to lumpectomy.

☐ Deny payment by all payers for technologies that are not approved for Medicare coverage.

Utilization controls also have a role—a double role, in fact. Consumers who in the past were passive in care decisions are now growing more active in making choices. However, information is needed for informed choice. Health-benefit plans can assist in these choices by making information and alternatives readily available through such utilization controls as second surgical opinion, peer review, and quality-of-care findings.

Sources of Information for Technology Assessment

While there is no clearinghouse for information on technology assessment at present, Congress has requested that the Institute of Medicine of the National Academy of Sciences, with federal and private-sector financing, develop a center to act as a source of such information.

A number of ongoing technology assessment activities provide information on the status, safety, and efficacy of a technology and provide data on peer judgment and/or studies of effectiveness. Little systematic analysis is conducted on cost effectiveness or on appropriateness of use of the technologies, except for isolated special studies. Only rarely are the social and ethical issues systematically raised even in formal technology assessments.

The following are the main resources for technology assessment:

National Center for Health Services Research and Technology Assessment (NCHSRTA). Provides advice to the Health-Care Financing Administration (HCFA) on Medicare coverage issues. Most of this advice is based on synthesis of the literature, and consultation with the National Institutes of Health, Food and Drug Administration, Centers for Disease Control, and specialty societies. The action that HCFA takes is made public, as are NCHSRTA's recommendations.

Food and Drug Administration (FDA). Assesses drugs and devices for safety and efficacy but does not assess procedures or medical techniques other than drugs and devices. For example, FDA assessed the balloon catheter but not the angioplasty technique.

National Institutes of Health (NIH). Funds and conducts clinical trials on safety and efficacy of medical-care procedures and techniques. The findings are published in medical literature. Also conducts consensus conferences on scientific technology issues and occasionally, with NCHSRTA, will address cost and social issues. The findings are widely publicized in the media and in medical journals.

American College of Physicians. This organization has established a major technology assessment activity and assists certain organizations like Blue Cross in providing information on the effectiveness of technology. Its findings led Blue Cross to discontinue payment for the large battery of tests that were automatically provided on hospital admission.

American Medical Association (AMA). The AMA has established a review group to provide practicing physicians with information on the safety and efficacy of medical technology. Findings are published in the *Journal of the American Medical Association*.

Office of Technology Assessment, U.S. Congress (OTA). Conducts periodic studies on specific technologies and assessment methodology.

Study Groups. There are several small study groups that conduct ad hoc assessments or studies of cost effectiveness—for example, those sponsored by the Rand Corporation, Harvard, and Dartmouth.

Assessment as a Means of Influencing Physicians

As a spinoff of the Wennberg small-area studies of variations in medical practice, the Maine Medical Assessment Program seeks to use the findings of these studies to feed back and influence physicians' practice patterns. The objectives of the program follow:

1. Develop an institutionalized program, sponsored and supported by organized medicine, that deals effectively with scientific, behavioral, and social issues associated with the practice variation.

2. Evaluate the outcome of two strategies to improve clinical decision making: (1) move atypical practice styles closer to consensus; and (2) resolve issues related to lack of consensus regarding disputed treatments.

When the National Center for Health Care Technology existed, it established links to Professional Standards Review Organizations to diffuse findings of technology assessment into medical practice.

The NIH consensus development conferences are designed to influence leading practitioners and community physicians by developing peer consensus on technology.

Through health-benefit plans and coalitions, the private sector can play a major role in stimulating the expansion of technology assessment activities in a number of ways, including pressures to expand federal activity related to Medicare coverage issues and support of clinical trial activity; financial assistance to technology assessment centers at universities or to groups such as the Institute of Medicine; active involvement in voluntary and mandatory planning and capital decisions; programs to review utilization, appropriateness, and quality; and reimbursement policies designed to stimulate appropriate and cost-effective use of technology.

Conclusions

Redesign of health benefits, utilization controls, prospective payment, and management of health benefits will have only limited short-term effects on health-care costs unless a rational and coherent approach to technology and its diffusion is developed. When resources are finite, choices have to be made. The need to make hard choices involving life-and-death issues can be reduced if resources are not used for unsafe, ineffective, inappropriate, marginal, overcapitalized, and underutilized technologies and if cost effectiveness is considered. Management and labor need to incorporate technology assessment and diffusion issues as integral parts of health-benefits design and management.

RECOMMENDATIONS

26 Employers and unions, working through coalitions and local health-planning agencies, should press for an area plan that establishes where high-cost technological services are to be sited and should also make a compact to abide by the plan. The plan should be designed to meet that area's needs efficiently, taking due account of accessibility, quality, and cost.

27 Employers and unions should ensure that managers, employees, dependents, and retirees are fully informed about the area plan.

28 Employers and unions should ensure that their health-benefits plans operate in the following ways:
— High-cost technological services are reimbursed only if they are performed at facilities designated by the area plan. If there is no area plan, facilities should be designated on the basis of criteria of quality and volume.
— The charges allowed for high-cost technological services are reviewed annually to keep them in line with (a) the normal decline of actual costs, and (b) the greater skill of physicians who perform the services and their increased number.
— Payment is denied for any technology not approved for Medicare coverage.

NOTES

1. U.S. Congress, House of Representatives, 95th Congress, 2nd Session Report #95-1190 by the Committee on Interstate and Foreign Commerce, Health Services Research, *Health Statistics and Health Care Technology Act* (Washington, D.C.: U.S. Government Printing Office, 1978).

REFERENCES

Gay, James J., and Jacobs, Barbara Sax, eds. *The Technology Explosion in Medical Science: Implications for the Health Care Industry and the Public (1981-2001)*. Jamaica, N.Y.: SP Medical and Scientific Books, 1983.

Hanft, Ruth S., and Eichenholz, Joseph. "The Regulation of Health Technology." *Annals of the American Academy of Political and Social Sciences*, Fall 1979.

Hanley, Daniel F. "The Maine Medical Assessment Program: Informed Inquiry by Practicing Maine Physicians into Common Medical and Surgical Treatments." An unpublished paper commissioned by Work in America Institute, January 9, 1985.

U.S. Department of Health, Education and Welfare. *Medical Technology: The Culprit Behind Health Care Costs.* Proceedings of the 1977 Sun Valley Forum on National Health. DHEW Publication No. 79-3216. Washington, D.C.: U.S. Department of Health, Education and Welfare, 1979.

GLOSSARY

Blue Cross/Blue Shield — A not-for-profit insurer of health benefits.

Capitation — A prospective flat sum per person or family paid periodically to a provider for a defined set of benefits; not a per-service payment, and not directly related to the actual use of services by an individual.

Case management — Coordination of the total care for the patient or family, usually provided by a physician or nurse.

Certificate of need — Approval to construct or renovate healthcare facilities or purchase equipment by a state agency before a hospital can proceed with construction, renovation, or acquisition of equipment. Reimbursement for capital costs can be denied if certificate-of-need approval is not granted. In a few states, certificate-of-need approval is required for capital expansion of physicians' offices or free-standing facilities.

Charge — A price set by the provider for a specified service, usually above the cost of the service.

Claim — A form submitted for payment of health expenditures by a provider or consumer that identifies the enrollee; the patient; demographic, diagnostic, and treatment information; and costs or charges.

Coinsurance — A percentage of the bill paid by the consumer, most often 10 percent or 20 percent.

Commercial insurance company — A for-profit mutual or stock company that sells health and other insurance.

Concurrent review — Ongoing review of a medical treatment plan, usually for hospitalized patients.

Copayment — A flat, nominal sum of money paid by the consumer for a service, for example, $2 per visit or prescription.

Cost reimbursement — Payment based on the actual cost of providing the service.

Cost sharing — An amount paid by the consumer directly; cost sharing includes premiums, deductibles, coinsurance, and copayments.

Deductibles — A dollar sum, usually $100-$200, that must be paid before health-benefit payments begin.

Diagnostic related group (DRG) — A method of prospective payment based on average costs for 468 different diagnostic groups.

Employee assistance programs (EAPs) — Programs designed to assist workers and their families with such problems as alcoholism, drug abuse, mental illness, and legal matters; the programs' objectives are to reduce absenteeism and restore productivity.

Fee for service — A charge rendered by the provider for a discrete service.

Health maintenance organization (HMO) — An organization that directly provides or contracts for all inpatient and outpatient services for enrollees at a prospectively set capitation rate. The enrollee must use the providers who are part of the group or plan, or pay out of pocket for their care. There are two forms: the individual practice association (IPA) and the prepaid group practice (PGP).

Medicaid — A federal/state matching welfare program that pays health benefits for public-assistance recipients; for those who are medically indigent and are aged, blind, disabled; or for families with dependent children.

Medically indigent — Individuals who do not receive public assistance, or those who have incomes below a specified amount and who have either inadequate health insurance or none.

Medicare — A federal health insurance program that pays health benefits for eligible people over 65, for those who are deemed permanently and totally disabled, or for those who have end-stage renal disease.

Peer review — Review of utilization and quality performed by health professionals (mainly physicians and nurses).

Preadmission certification — Prior approval for the receipt of medical-care services—most commonly for hospital and nursing-home stays, elective surgery, and expensive diagnostic procedures.

Preferred provider organization (PPO) — A new type of provider organization that arranges for services with selected hospitals and physicians on the basis of predetermined charges or fees.

Premiums — Periodic payments by employers or consumers for health-benefits coverage.

Prevailing fee — The average or a percentile of the range of fees charged for a given service by the providers in a medical-service area.

Primary-care physician — General or family practitioner, pediatrician, internist, obstetrician, or gynecologist.

Professional review organization (PRO) — Peer review of health-care services by physicians and nurses to determine necessity and appropriateness of care ordered and/or provided.

Prospective payments — Payments set in advance of use of services. Prospective payments can be based on historic costs, budgets, average costs for diagnostic related groups, percentiles, or discounts from normal charges.

Quality assurance — Covers a wide range of programs to assure that providers of service meet predetermined criteria for training and certification, standards for facilities, morbidity and mortality reviews, infection control, and other quality-of-care measures.

Reinsurance — An insurance policy that underwrites catastrophic costs or costs beyond a defined level.

Retrospective cost or charge reimbursement — Payment of actual costs or charges after they have been incurred. The provider controls the costs or charges.

Second surgical opinion — A plan that pays for a second opinion related to the need for an elective surgical procedure.

Self-insurance — The type of insurance in which the company or joint labor-management welfare or health fund underwrites (assumes the risk for) the financing of health benefits.

Stop loss — A cap on the amount of cost sharing a subscriber to a health plan is required to pay, that is, the carrier pays 100 percent of costs beyond a stated amount; sometimes referred to as catastrophe insurance.

Technology assessment — Determination of the safety, efficacy, and cost effectiveness of drugs, medical devices, and diagnostic and treatment procedures as well as social and ethical issues related to them.

Third-party payer — The payment agent for a person insured for health expenditures or an enrolled member of a health-benefits plan. This term usually refers to private health insurance companies, Medicare, or Medicaid, but also includes self-insured plans.

Usual, customary, and reasonable (UCR) charges — The fee established by an individual practitioner (physician) that is charged to the majority of patients.

Utilization review — Review of utilization of health services against preestablished norms to assure appropriateness of use of services, sites of services, and admissions and lengths of stay in institutions.

Wellness programs — Programs designed to prevent illness and maintain or improve health status. Includes a wide range of occupational health and safety, disease screening and prevention, smoking, alcohol and drug abuse, employee assistance, exercise, and nutrition programs.

INDEX

ABGWIU. *See* Aluminum, Brick, and Glass Workers International Union
Accreditation of institutions, 54; and quality of care, 66
Acute care, 47
AFL-CIO: employer-union coalitions and, 21, 22; health-care cost-containment techniques, 18; as member of Group of Six, 19-20
ALCOA. *See* Aluminum Company of America
Aluminum, Brick, and Glass Workers International Union (ABGWIU): employer-union joint action and, 30-31
Aluminum Company of America (ALCOA): cost sharing and, 48; employer-union joint action and, 30-31
Aluminum Workers of America: cost sharing, 48
AMA. *See* American Medical Association
American Cancer Society: wellness programs, 88
American College of Physicians, 147
American Hospital Association: data collection, 63; as member of Group of Six, 19-20; supply data, 60
American Medical Association (AMA), 147-148; medical education funding and, 3; as member of Group of Six, 19-20
American Telephone and Telegraph Company (AT&T): design of health-care benefits, 34; employer-union joint action and, 28-30

"An Appropriate Role for Corporations in Health-Care Cost Management," 114
"Appropriateness Evaluation Protocol," 64, 65
Area resource file, 60
Arthur D. Little, Inc., 132
AT&T. *See* American Telephone and Telegraph Company

Baltimore Gas and Electric Company: wellness program and, 81
Bank of America: financial incentives, 34; preadmission certification, 64
Beneficiaries: cost sharing, 47-49
Big Steel. *See* United Steelworkers of America
Birthing Centers, 9; as hospital alternative, 44, 45, 46
Blue Cross/Blue Shield Association, 151; claims and utilization data, 62; employer-union coalitions and, 21, 22; evolution of health benefits, 35; HMOs and, 96, 98; history, 38; hospital capacity and, 130; as member of Group of Six, 20; payment for home health services, 46
Business Roundtable, The: approach to health initiatives, 53; health-care costs and, 49-50; health-care strategies of, 114-116; Health Initiatives project and, 12; as member of Group of Six, 20; Task Force on National Health, 12

California Medicaid program, 97

California Psychological Health Plan: HMOs and, 94
Campbell Soup Company: wellness program, 81
Caper, Philip, 68
Capital Area Coalition, 21
Capitation, 93-94; definition, 151; health-care funding and, 3; HMOs and, 95-96
Case management: definition, 151
Caterpillar Tractor Company: data collection, 63; employer-union involvement in HMOs, 103
Centers for Disease Control, 147
Certificate of need: definition, 151
Certification, 66
Charge: definition, 151
Chrysler Corporation: HMO education and, 106
CIBA-GEIGY Corporation: health-care strategies, 119-121
CIGNA Corporation: design of health-care benefits, 38; employer-union coalitions and, 21; HMOs and, 96
CINCH. *See* Comprehensive Insurance Claims Handling
City Federal Savings and Loan Association: wellness program, 81
Claim: definition, 151
Claims data, 55-56, 60
Claims review, 54-55, 60
Closure: hospital capacity and, 138
Coinsurance: cost sharing and, 49; definition, 151
COLA. *See* Cost of living allowance
Columbia University School of Public Health and Administrative Medicine, 126
Commercial insurance company: definition, 152
Communications Workers of America (CWA): design of health-care benefits, 34; employer-union coalitions and, 22; employer-union joint action and, 28-30
Comprehensive Insurance Claims Handling (CINCH), 61
Concurrent review: definition, 152; utilization control, 62

Consolidation: hospital capacity and, 138
Control Data: wellness program, 81
Coors: wellness program, 81
Copayment: cost sharing and, 49; definition, 152
Corporate Council, 122
Corporate culture, 76, 78, 80
Corporate Health Strategies, 62
Cosharing: employer-union coalitions and, 22
Costs: control in wellness programs, 78; hospital capacity and, 129
Cost of living allowance (COLA), 29-30
Cost reimbursement: definition, 152
Cost sharing, 25; beneficiaries and, 47-49; coinsurance and, 49; copayment and, 49; deductibles and, 49; definition, 152; employer-union coalitions and, 23; health-care management and, 56; HMOs and, 100; premiums and, 49
CPT-4, 61, 69
Credentials, 54
Current Procedural Terminology, 61
CWA. *See* Communications Workers of America

Dartmouth Medical School, 40, 62, 100; cost-effectiveness studies, 148
Data: claims, 115; collection problems, 70-71; cost, 71; epidemiological, 68; morbidity, 67; mortality, 67; program, 60; quality, 71; use, 71
Deductibles: cost sharing and, 49; definition, 152
Deere and Company: claims processing systems, 61; employer-union involvement in HMOs, 103; HMO education, 106; preadmission certification and, 64
Department of Health and Human Services, 60
Diagnostic centers, 9; as hospital alternative, 45
Diagnostic related group (DRG): claims and utilization review and,

61; cost analysis, 69; definition, 152; employer-union coalitions and, 22; payment procedures of, 7; reimbursement incentives, 8
Disability: costs, 117
DRG. *See* Diagnostic related group
Dunlop, John T., 19
Du Pont. *See* E.I. Du Pont de Nemours

EAP. *See* Employee assistance programs
Eastman Kodak Company: HMOs and, 102
Economic Alliance for Michigan, 20, 132
Education: foreign medical, 66-67; health-care plan management, 55
E.I. Du Pont de Nemours: data collection, 63
Ellwood, Paul, 94-95
Emergency centers, 41
Employees: morale of, 78; wellness programs for, 78
Employee-assistance program (EAP): definition, 152; work-site wellness program, 76, 79
Employers: health-care costs and, 83; health-care strategies of, 113-128; hospital capacity and, 136-137; and providers, 107; role in health care, 13; involvement with unions in HMOs, 102-104; and union joint action, 26-31; commitment to wellness programs, 87
Ergonomics: injury at work, 117; work-site wellness programs, 80
Experience rating, 38
Extended-care facilities, 43-44; as hospital alternative, 46; *see also* Intermediate care facilities; Nursing homes

Fairfield-Westchester coalition, 62
FDA. *See* Food and Drug Administration
Fee for service: definition, 152
Fielding, Jonathan, 83
Flexner, Abraham, 2
Flexnerian reform, 3

Flexner report, 35
Food and Drug Administration (FDA), 146, 147
Ford Motor Company, 20; employer-union involvement in HMOs, 103; HMO education and, 106; wellness program, 81
Free-standing diagnostic centers, 42

General Accounting Office: HMOs and, 97
General Electric Company: health-care research and, 12; health-care strategies, 122-123
General Mills, Inc.: employer-union involvement in HMOs, 103
General Motors Corporation: design of health-care benefits, 34; employer-union joint action, 27-28; joint action procedures, 31-32
GMENAC. *See* Graduate Medical Education National Advisory Committee Report
Governor of New York: HMO education program, 106
Graduate Medical Education National Advisory Committee Report (GMENAC), 4
Group Health of Puget Sound: HMOs and, 95
Group Health of Washington: HMOs and, 95
Group of Six, 19-20; recommendations, 20

Harvard Medical School: cost-effectiveness studies, 148
HCFA. *See* Health Care Financing Administration
Health care: cost-benefit methodologies, 82; cost negotiations, 57; costs, 5-8, 49-50; cost sharing, 50; demographics, 38, 55-56, 61; evaluation, 58; expenditures, 5; funding, 5; future of, 11-12; guidelines, 13-15; health improvement and, 78; history, 1-8; hospital alternatives to, 43-44; incentives, 50;

management model, 58-59; monitoring, 58; nonhospital services, 39-40; objectives, 14-15; options, 50; overlapping coverage, 51; plan design and communication, 58; private sector and, 9-11; program data, 60; promotion of, 115; quality of care, 14; reimbursable services, 45; scope of benefits, 50; selection policies for management of, 54-55; selection, 58; technology, 37, 47, 142-150
Health Care Financing Administration (HCFA), 63, 147; HMOs and, 96-97
Health Care Management Team, 122
Health Care Network, 122
Health Care Policy Department, 121
Health Data Institute, 62
Health insurance: private, 2
Health Insurance Association of America, 62; as member of Group of Six, 20
Health Interview Surveys, 63
Health maintenance organization (HMO), 93-111; accessibility, 105-106; competition, 101-102; definition, 93, 95-96, 152; dental coverage, 28, 94; design of health-care benefits, 36, 39; development, 107; education concerning, 71; employee enrollment, 106-107; employer pros and cons, 104-105; employer-union involvement, 102-104; establishment of, 23; future of, 108-109; growth of, 106-107; health-care management and, 56; history, 96-98; management of health-care benefits, 54; payment procedures, 8; performance, 98-101; physician recruitment into, 9; promotion of enrollment, 20-21; and providers, 107; union recommendations, 19. *See also* Preferred provider organization
Health Policy Corporation, 62
Health Systems Agencies (HSAs), 115; hospital capacity and, 135
Hewlett-Packard Company: employer-union involvement in HMOs, 103; HMO education, 106-107
Hill-Burton Act, 3, 35
HMO. *See* Health maintenance organization
Home health care, 10, 30, 37; as hospital alternative, 44-45; reimbursement costs of, 43
Honeywell Inc.: employer-union involvement in HMOs, 103
Hospice, 9; as hospital alternative, 44
Hospital: for-profit, 10-11; payment procedures, 58
Hospital capacity reduction, 129-141; assistance for, 138-139; bed-reduction plans, 137-138; hazards, 133-136; planning process, 139
Hospital Corporation of America: preadmission certification, 64; wellness program and, 81
Hospital Discharge Surveys, 63
HSA. *See* Health Systems Agencies

IBEW. *See* International Brotherhood of Electrical Workers
IBM: HMO education and, 106; wellness programs, 80, 81
ICD-9-CM, 61, 69
ILGWU. *See* International Ladies' Garment Workers' Union
Independent practice association (IPA): hospital capacity and, 130; HMOs and, 94, 95; HMO performance and, 98-99
Informed Choice Plan, 27
Institute of Medicine, 146-147
Intermediate care facility, 43-44; as hospital alternative, 46. *See also* Extended-care facility; Nursing homes
International Brotherhood of Electrical Workers (IBEW): design of health-care benefits, 34; employer-union joint action and, 28-30
International Classification of Diseases, Modified, 61
International Ladies' Garment Workers' Union: design of health-care benefits, 36; employer-union

involvement in HMOs, 102; health-care costs and, 23; health-care strategies, 123, 125-126
International Typographical Union, 21
Iowa Business-Labor Coalition on Health, 22
IPA. *See* Independent practice association

JCAH. *See* Joint Commission for the Accreditation of Hospitals
John Hancock Mutual Life Insurance Company: HMOs and, 96
Johnson & Johnson: wellness program, 81
Joint Commission for the Accreditation of Hospitals (JCAH), 66: design of health-care benefits, 36
Joint Labor-Management Committee of the Retail Food Industry Trust, 20
Jones & Laughlin Steel Company: cost sharing, 48

p
Kaiser Aluminum & Chemical Corporation: HMOs and, 94; 96, 97-98; employer-union involvement in HMOs, 103; utilization controls, 64
Kaiser Foundation: employer-union involvement in HMOs, 103
Kaiser Plan, 94, 102
Kansas Employers Coalition on Health, 22
Kimberly-Clark Corporation: wellness programs, 81
Kodak. *See* Eastman Kodak

Labor-Management Group, 19
Levi Strauss & Company: preadmission certification requirements, 64
Lewin and Associates, 134
Lewin study, 134-136
Licensure, 54
"Live for Life Program," 81
Lockheed Corporation: employer-union involvement in HMOs, 103
Long Island Jewish Medical Center, 126

Luft, Harold, 98, 130

Maine Medical Assessment Program: cost-effectiveness studies, 148
Master Freight: employer-union joint action and, 29-30
Medicaid program, 5; definition, 152; history, 3
Medical-Care Use and Expenditures Surveys, 63
Medically indigent: definition, 152-153
Medical technology: use of, 142-150
Medicare program, 5; definition, 153; design of health-care benefits, 36; diagnostic related groups and, 8; history, 3; home health services and, 37; renal dialysis centers and, 42; severity indicator, 63
Merger: hospital capacity and, 138
Metropolitan Life Insurance Company, 62; design of health-care benefits, 38; health-care research and, 12; HMOs and, 96; wellness program, 81
Midwest Business Group on Health: data collection, 62; design of health-care benefits, 34; health-care management model, 58-59
Milwaukee Association of Commerce, 137
Minneapolis Foundation for Medical Care Evaluation, 65
Minnesota Coalition on Health Care Cost, 21, 62
Mitre Corporation: preadmission certification, 64

National Academy of Sciences, 147
National Ambulatory Medical-Care Surveys, 63
National Center for Health Care Technology, 148; technology assessment, 143
National Center for Health Services Research and Technology Assessment (NCHSRTA), 143, 146, 147
National Center for Health Services Research: and HMO performance, 98-100

National Center for Health Statistics (NCHS), 67; data collection, 63
National Health Expenditures, 63
National Institutes of Health (NIH), 147, 148; with Food and Drug Administration, 147; medical technology use and, 144
NCHS. See National Center for Health Statistics
NCHSRTA. See National Center for Health Services Research and Technology Assessment
New York Telephone Company: wellness programs, 81
NIH. See National Institutes of Health
Nixon Administration: HMOs and, 94
Nonprofit organizations: wellness programs, 88
Nursing homes: as hospital alternatives, 43-44, 46. See also Extended care facilities; Intermediate care facilities

Office of Health Maintenance Organizations: and employer attitudes toward HMOs, 104; HMOs and, 97-98
Office of Technology Assessment (OTA), U.S. Congress, 148
Oregon Public Employees Union: employer-union joint action and, 30
Orkand Corporation, 134
Orkand study, 134-136, 138-139
OTA. See Office of Technology Assessment, U.S. Congress
Owens-Illinois, Inc.: health-care strategies, 121-122

Palo Alto Clinic: HMOs and, 95
Payments: Capitation, 3, 93-94, 95-96, 151; fee-for-service, 2; hospital, 58; other provider, 58; physician, 58; prospective, 153; third-party, 2
Peer review; definition, 153
Pfizer Inc.: health-care costs, 50; health-care research and, 12; health-care strategies, 116-118; wellness programs, 78-79

PGP. See Prepaid group practice
Physician: payment, 58; primary-care, 153
Pioneer Hi Bred International, Inc.: wellness program, 81
PPO. See Preferred provider organization
Preadmission certification: definition, 153; for nonemergency services, 64
Preauthorization, 28
Preferred provider organization (PPO), 20-21; definition, 153; dental coverage with, 28; design of health-care benefits, 39; development, 10; education of, 71; employer-union coalitions and, 22; employer-union joint action, 27-28; health-care management, 56; management of health-care benefits, 54. See also Health maintenance organization
Premiums: cost sharing and, 49; definition, 153
Prepaid group practice (PGP): HMOs and, 95-96; HMO performance, 98-99; hospital capacity and, 130
Prevailing fee: definition, 153
Primary-care physician: definition, 153
Private sector: health-care development in, 9-11
PRO. See Professional review organization
Professional review organization (PRO): data collection, 62-63; definition, 153; and quality of care, 67
Professional standards review organization (PSRO), 115; cost-effectiveness studies, 148; data collection, 62-63; utilization control, 65
Profile analysis, 64
Prospective payments: definition, 153
Providers: and HMOs, 107
Prudent-buyer strategies, 55
Prudential Insurance Company: design of health-care benefits, 38; employer-union involvement in HMOs, 103; HMOs and, 96

PSRO. *See* Professional standards review organization

Quaker Oats Company: cost sharing, 48
Quality assurance, 55; definition, 153
Quality of care: 66-68

Rand Corporation, 24; cost-effectiveness studies, 148; HMOs and, 100
Rand experiment, 24-26
Rand study: cost sharing, 48; HMO performance and, 98; hospital capacity and, 131. *See also* Rand experiment
Reagan Administration, 43
Reinsurance: definition, 153
Renal dialysis centers, 42-43
Retirees Service Program, 28
Retrospective cost reimbursement: definition, 154
Reuther, Walter, 124
Review: of claims and utilization, 60
R.J. Reynolds Industries, Inc.: cost sharing, 48; employer-union involvement in HMOs, 103
Ross Loos Clinic: HMOs and, 94, 95

Screening. *See* Testing
Second surgical opinions, 28; definition, 154; utilization control, 65-66
SEIU. *See* Service Employees International Union
Self-insurance: definition, 154
"Sentinel effect," 64-65
Service Employees International Union (SEIU): cost-effectiveness programs of, 23; design of health-care benefits, 34; employer-union joint action and, 30; health-care programs, 19
Services: quality of care, 67
Social Security Amendments, 6, 40
South Central Bell Telephone Company: cost sharing, 49
Standard metropolitan statistical area, 60

Stop loss: definition, 154
Surgery: ambulatory, 28; inpatient, 28
Surgicenter, 9, 40; as hospital alternative, 45

"Takebacks," 23
Tax Equity and Fiscal Responsibility Act, 6
TCS. *See* Teamster Center Services
Teamster Center Program, 126
Teamster Center Services (TCS), 126
Teamsters: design of health-care benefits, 34; employer-union joint action and, 29-30; health-care recommendations, 19; health-care strategies, 126-127
Teamsters Joint Council No. 16 Hospitalization Trust Fund, 126
Technology: medical use of, 142-150
Testing: preadmission, 28, 49, 65; screening, 30
Texaco, Inc.: wellness program, 81
Third-party payer: definition, 154
Training: foreign, 66-67

UAW. *See* United Auto Workers
UCR. *See* Usual, customary, and reasonable charges
Unions: and employer involvement in HMOs, 102-104; joint actions with employers, 26-31; health-care management and, 18; health-care recommendations, 19; health-care strategies, 113-128; hospital capacity and, 136-137; role in health care, 13
United Auto Workers (UAW), 20; design of health-care benefits, 34; employer-union coalitions and, 21, 22; employer-union involvement in HMOs, 103; employer-union joint action, 27-28; health-care costs and, 23; health-care recommendations, 19; health-care strategies, 123, 124-125; HMOs and, 94, 98; HMO education, 106; joint action procedures, 31-32

U.S. Administrators, 62
U.S. Department of Health and Human Services: and Rand experiment, 24; wellness programs, 80, 81
United Steelworkers of America (USA): employer-union coalitions and, 22; employer-union joint action, 29
University of California, San Francisco: HMO performance, 100
Urgent-care centers, 9, 41; as hospital alternative, 45
USA. *See* United Steelworkers of America
Usual, customary, and reasonable (UCR) charges: definition, 154
Utah Health Cost Foundation, 62
Utilization: claims data, 60
Utilization control, 45, 55; mechanisms for, 63-66
Utilization management, 58
Utilization review, 28, 54-55, 60; definition, 154

Washington Business Group on Health: wellness programs, 80

Wellness programs, 75-92; behavior risks, 80; benefits quantification, 86-87; capacity assessment, 86; classification, 79-80; competition, 88; confidentiality, 87; contracting for services, 87; corporate commitment, 87; cost, 88; definition, 154; demonstration basis, 87; early detection in, 79; early intervention in, 79-80; eligibility for, 86; evaluation of plan design, 87; incentives, 88; liability risks, 88; needs assessment, 86; objectives, 85-86; practice assessment, 86; program criteria, 85-87; quality control of, 88; rationale for, 77-78; selection of, 84-85
Wennberg, John, 40, 62, 68, 100, 130, 136
Wennberg small-area statistical studies, 131, 148
Westinghouse Electric Corporation: preadmission certification, 64

Xerox Corporation: cost sharing, 48; HMOs and, 102; wellness programs, 80

NO LONGER THE PROPERTY OF THE UNIVERSITY OF R.I. LIBRARY